TIME

MANAGING EDITOR Richard Stengel
DESIGN DIRECTOR D.W. Pine
DIRECTOR OF PHOTOGRAPHY Kira Pollack

Your Body
The Science of Keeping It Healthy

EDITORS Lisa Davis, Lorie Parch
DESIGNER Sharon Okamoto
PHOTO EDITOR Dot McMahon
WRITERS David Bjerklie, John Cloud, Stacey Colino, Kristina Dell, Christine Gorman, Jeffrey Kluger, Kristin Koch, Michael D. Lemonick, Catherine Mayer, Harry McCracken, Colleen Moriarty, Myatt Murphy, Regina Nuzzo, Mehmet Oz, M.D., Alice Park, Julia Savacool, Alexandra Sifferlin, Joel Stein, Maia Szalavitz, Bryan Walsh, Martha C. White
REPORTERS Lina Lofaro, Damien Scott, Jenisha Watts
COPY EDITOR David Olivenbaum
EDITORIAL PRODUCTION David Sloan

TIME HOME ENTERTAINMENT

PUBLISHER Jim Childs
VICE PRESIDENT, BRAND & DIGITAL STRATEGY Steven Sandonato
EXECUTIVE DIRECTOR, MARKETING SERVICES Carol Pittard
EXECUTIVE DIRECTOR, RETAIL & SPECIAL SALES Tom Mifsud
EXECUTIVE PUBLISHING DIRECTOR Joy Butts
DIRECTOR, BOOKAZINE DEVELOPMENT & MARKETING Laura Adam
FINANCE DIRECTOR Glenn Buonocore
ASSOCIATE PUBLISHING DIRECTOR Megan Pearlman
ASSOCIATE GENERAL COUNSEL Helen Wan
ASSISTANT DIRECTOR, SPECIAL SALES Ilene Schreider
BRAND MANAGER Bryan Christian
ASSOCIATE PRODUCTION MANAGER Kimberly Marshall
ASSOCIATE BRAND MANAGER Isata Yansaneh
ASSOCIATE PREPRESS MANAGER Alex Voznesenskiy

EDITORIAL DIRECTOR Stephen Koepp
COPY CHIEF Rina Bander
DESIGN MANAGER Anne-Michelle Gallero

SPECIAL THANKS TO: Katherine Barnet, Jeremy Biloon, Susan Chodakiewicz, Rose Cirrincione, Jacqueline Fitzgerald, Christine Font, Jenna Goldberg, Hillary Hirsch, David Kahn, Amy Mangus, Nina Mistry, Dave Rozzelle, Ricardo Santiago, Adriana Tierno, Vanessa Wu, Time Inc. Premedia

---

ISBN 10: 1-61893-083-4
ISBN 13: 978-1-61893-083-5
Library of Congress Control Number: 2013938648

We welcome your comments and suggestions about TIME Books. Please write to us at:
TIME Books, Attention: Book Editors, P.O. Box 11016, Des Moines, IA 50336-1016.
If you would like to order any of our hardcover Collector's Edition books, please call us at:
1-800-327-6388, Monday through Friday, 7 a.m. to 8 p.m., or Saturday, 7 a.m. to 6 p.m., Central Time.

Some articles in this book have appeared previously in TIME, HEALTH, and REAL SIMPLE magazines or their websites, Time.com, Health.com, and RealSimple.com.

Contents

Living Long and Living Well

It's not so hard—and it's not all genes. Here's what we can do today to get some extra tomorrows.

BY DR. MEHMET OZ

WHEN EXPLORER AND LON-gevity investigator Dan Buettner guided me into the Costa Rican rainforest a few years ago, the cente-narians I met there greet-ed me with the customary "Pura vida"—variously translated as "Pure life," "Full of life" or even "This is liv-ing!" These vibrant people are indeed living the pure life. We'd all do well to learn their secrets.

Although our genes help determine everything from our height to our risk of heart disease, we're making a monumental mistake if we assume we can't influence those genes, especially when it comes to aging. Science is uncovering miraculous biological processes that con-trol why we age the way we do, piling up evidence that even unwanted genes can work in our favor—or at least do us less harm.

Indeed, there's no reason we can't live to 100, with energy and good health. Here's why: longevity is not really about preventing disease. After all, getting rid of heart disease and cancer gains us, on average, less than a decade of life. And if we lived those extra years strug-gling with frailty, what would we have gained? No, the real goal isn't to avoid illness or breakdown. The goal is to recover faster and better.

Identifying solutions will require decades, in part be-cause it takes 30 years of research to determine whether taking a pill for 20 years will add a decade of life. So here are some reasonable steps I've offered my own family, culled from studying long-living populations around the world and cutting-edge scientific research.

Daily rigorous physical activity not only helps strengthen bones and the heart but also teaches balance, critical in preventing the falls that are a leading cause of death as we age. The ability to exercise remains the single most powerful predictor of longevity. If you can't walk a quarter-mile in five minutes, your chance of dying within three years is 30% greater than that of faster walkers.

Humans are designed to be physically active through-out their lives, so shoot for at least three 30-minute work-outs weekly—and break a sweat. Add a half-hour per week of weight-lifting and another half-hour of stretching.

Get 15 minutes of sun every day (or take 1,000 IU of vitamin D), and take 1,000 mg of calcium. Supplement the calcium with 500 mg of magnesium to avoid consti-pation. All of this will help promote bone strength as you exercise. Costa Ricans get these benefits naturally: they're exposed to lots of sun, which keeps their vitamin D levels high; they drink hard, mineral-rich water and eat a traditional diet rich in calcium.

In the U.S., we're not so lucky. Insufficient vitamin D is our most important vitamin deficiency and is possibly a factor in our high levels of cancer, autoimmune ailments and heart disease. If you live north of a line between At-lanta and Los Angeles, the winter sun is probably too weak to deliver enough light, so you'll need supplements. And hard water isn't found in all parts of the country.

Choose foods that look the same when you eat them as when they come out of the ground. The phytochemi-cals and micronutrients in whole foods (ones without food labels) support the natural rejuvenating processes of the body.

Obese people, in whom such processes become com-promised, tend to die younger in part because of systemic inflammation that occurs as a result of their weight. That leads to raised blood sugar, lousy LDL-cholesterol levels and high blood pressure, which all damage the lining of our arteries. The fat also increases cancer rates and leads to joint pain that limits physical activity. Automate your meal choices to create routines that make it easy to eat the right foods. Snacking on healthy foods every few hours helps you avoid hunger and the associated overeating.

Sleep more than seven hours a day. Sleep increases your levels of growth hormone, a critical vitality booster. Half of mature Americans have difficulty sleeping, and all of them may pay a longevity penalty. Try some simple sleep hygiene like dimming the lights 15 minutes before bedtime to stimulate melatonin.

Finally, have a purpose—your family, your work, your community. There may be no better longevity booster than wanting to be here. You have one life; it makes sense to love living it.

Feeling Fine

THE POWER OF MOOD
A PRIMER FOR PESSIMISTS
JUST SAY OM
CAN POSTURE CHANGE YOUR MIND?

NO ONE WHO'S SEEN THE TOLL TAKEN BY STRESS and sorrow on a friend or relative can doubt the physical consequences of thoughts and feelings. Still, it can be hard to grasp the breadth and depth of the mind's influence over the workings of the body. More and more research shows that physical health is intimately tied to mental well-being, and vice versa. In fact, the judicious tweaking of your psyche may be one of the best ways to speed recovery from disease or perhaps help prevent illness in the first place. Knowing this may make the maintenance of a light mood feel like a heavy burden. But science is also revealing effective and surprisingly uncomplicated ways to give your mind a tune-up. It seems that one way to change your brain is to move your body.

The Power of Mood

Depression can shrink your brain and shorten your life, while happiness is a tonic—literally. So what are you going to do about it?

✳ BY MICHAEL D. LEMONICK

T WASN'T ALL THAT LONG AGO THAT HEALTH EXPERTS CONSIDERED the mind and the body to be pretty much unrelated. They knew, of course, that the mind—our thoughts, memories, feelings, and consciousness itself—resides in the brain. They also knew that these processes are rooted in biology. Our sense of self, and everything that comes together to create it, is ultimately an interplay between electrical impulses and biochemical reactions. But it was long an article of faith in the medical world that the brain and body operated in parallel, each tending to its own anatomical business.

No more, however. Over the past few decades, doctors of the body and mind have come to realize that the physical realm and the mental one affect each other profoundly. Physical illness often leads to mental imbalance, and patients suffering from psychological ills seem especially vulnerable to serious physical disorders. But the spiral doesn't go only in a negative direction: a state of optimism and mental health can make the body healthier, and a healthy body can elevate the mind.

All this makes sense. The brain, after all, is just another organ, and it operates on the same biochemical principles as the thyroid or the spleen. It shouldn't come as a surprise that it can affect, and be affected by, other parts of the body, for good or ill. But the effect is potentially so significant that the medical profession has begun to focus serious attention and resources on trying to understand what's going on.

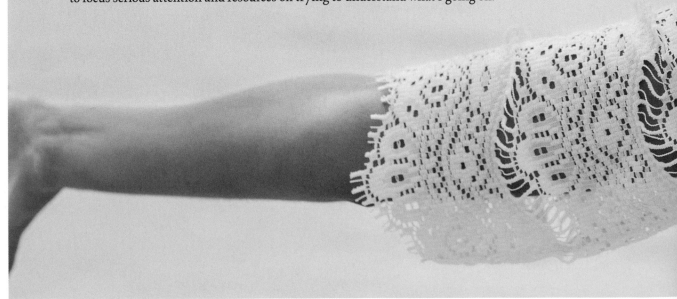

Depression may be the best-studied example of this interplay. It's well known that depression leads to suicide—tens of thousands each year in the U.S. alone. But depression's physical toll goes far beyond that, and even beyond the impact on depressed people's relationships and productivity (which costs the U.S. economy some $50 billion a year). Cardiologists have long noted that patients undergoing cardiac bypass and other kinds of open-heart surgery tend to come down with depression afterward, for example. And if you have had heart failure, you increase your risk of being readmitted to the hospital if you suffer from depression.

It's not simply that people tend to be depressed because they have a life-threatening illness, or that depressed people smoke, are too lethargic to take their medicine, or aren't motivated to eat right or exercise. "Even when we take those factors into consideration," says Dr. Dwight Evans, chairman of psychiatry at the University of Pennsylvania, "depression jumps out as an independent risk factor for heart disease. It may be as bad as cholesterol."

Heart disease is merely one in a long list of illnesses that worsen with depression. People with such afflictions as cancer, diabetes, epilepsy and osteoporosis all appear to run a higher risk of disability or premature death when they are clinically depressed. Fortunately, scientists have made great strides in sorting out the underlying causes of depression. It is almost certainly a defect in some combination of key genes, plus the right triggering environment. And researchers are well along in developing promising therapies, pharmacological and otherwise, to supplement what is already available. But while the disease-depression connection is becoming more and more clear, how to uncouple them is an uncharted process. "You would think that treatment would alter the negative relationship between depression and other illness," says Dr. Dennis Charney, dean of the Icahn School of Medicine at Mount Sinai in New York and former director of mood-and-anxiety-disorder intramural research at the National Institute of Mental Health (NIMH). But, he adds with proper scientific caution, "we don't have proof of that yet."

The idea that treating depression might lessen the severity of other diseases, though, makes basic biochemical sense. Everyday experience makes it clear that brain chemistry governs more than just the emotions. When your mind feels terror, the resulting surge of adrenaline makes your stomach churn. When your mind is sexually aroused, the body responds in unmistakable fashion. The effect is even more direct with the 60 or so chemicals known as neurotransmitters, which signal one cell that its neighbor has just sparked and that it should pass along the message. Brain chemicals such as serotonin circulate everywhere, not only in the brain. "Depression really is a systemic disorder," says Evans, "and many of the neurotransmitters that we believe are involved in the pathophysiology of depression have effects throughout the body."

Precisely how these powerful chemicals affect the course of heart disease, cancer, and other illnesses isn't yet well understood, but preliminary research has yielded tantalizing clues. When lots of serotonin circulates in the bloodstream, for example, it appears to make platelets less sticky and thus less likely to clump together in artery-blocking blood clots. For years, heart-attack survivors have been advised to take a children's aspirin daily for clot prevention; drugs like Prozac, which keeps serotonin in circulation, seem to have a similar effect.

Another mechanism may also be at work. It turns out that the heartbeat of a person with depression is unusually steady. That's not necessarily a good thing,

One study tracked **1,300** men for 10 years and found that those who called themselves optimistic had just half the heart-disease rates of those who didn't.

says Charney. "Ideally, your heart rate should be variable—it means your heart can respond appropriately to the different tasks it's called upon to respond to." Yet another possible link between heart disease and depression is a chemical called C-reactive protein (CRP). The liver normally produces CRP in response to an immune-system alarm when the body is infected or injured, and CRP is associated with the inflammation that results. For reasons still unknown, though, a 2003 study of depressed individuals found elevated levels of CRP. And in patients whose arteries have been damaged by the buildup of cholesterol plaques, heightened inflammation may increase the chance that a bit of plaque will break off and shut down an artery.

Diabetes is another illness that doesn't go well with depression. It's known that 10% of diabetic men and 20% of diabetic women also have depression—about twice the rate in the general population. It's natural to be depressed about having a chronic, potentially fatal illness, but that doesn't entirely explain the discrepancy. Moreover, depressed diabetics are much more likely than those without depression to suffer complications including heart disease, nerve damage and blindness. Somehow depression makes the body less responsive to insulin, the hormone that processes blood sugar—possibly through the action of cortisol, a hormone that can interfere with insulin sensitivity and that is often elevated in depressed patients.

Cortisol may also make depressed patients more prone to osteoporosis. Studies by Dr. Philip Gold and Dr. Giovanni Cizza at the NIMH have shown that premenopausal women who are depressed have a much higher rate of bone loss than their nondepressed counterparts, and this disparity increases as women pass through menopause. Cizza estimates that some 389,000 women get osteoporosis each year because of depression. Cortisol appears to interfere with the ability of the bones to absorb calcium. (Another class of chemicals, the pro-inflammatory cytokines, have also been implicated in osteoporosis and diabetes, but their role is less clear.)

Studies have established links between the incidence of depression and several other diseases, including cancer, Parkinson's, epilepsy, stroke and Alzheimer's. In some cases at least, researchers have clues, if not definitive evidence, as to which molecules might be involved. In Parkinson's, the problem is the death of cells in the brain that produce the neurotransmitter dopamine. Dopamine is crucial to the control of movement, but it's probably a major factor in mood as well.

"Depression almost certainly has multiple causes that produce similar symptoms," observes Dr. Bruce Cohen, president of McLean Hospital in Belmont, Mass.

That could explain why drugs that improve serotonin chemistry don't always work on depression—and why Parkinson's and depression can feed on each other. Epilepsy, stroke and Alzheimer's, which, like Parkinson's, involve physical alteration of the brain, probably also affect that organ's ability to make or process neurotransmitters—not only serotonin and dopamine but also the brain chemicals glutamate and norepinephrine, all of which may be involved in different forms of depression.

As Charney has noted, it hasn't been proved, in a rigorous, scientific sense, that treating depression will reduce the excess risks of complication or death from a coexisting illness. But if depression treatments rebalance the underlying biochemistry that worsens disease, there is every reason to expect that they will reduce its deadly impact. Research has also provided more persuasive reasons to feel re-

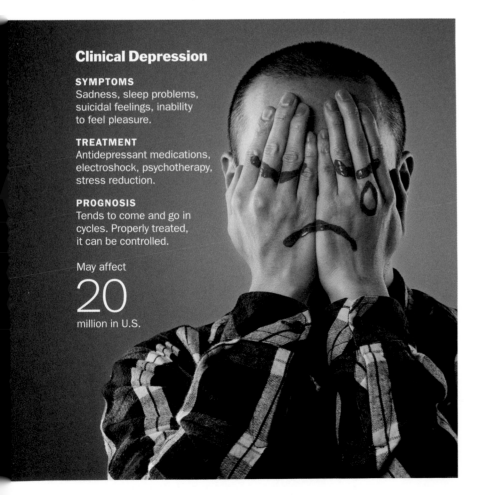

Clinical Depression

SYMPTOMS
Sadness, sleep problems, suicidal feelings, inability to feel pleasure.

TREATMENT
Antidepressant medications, electroshock, psychotherapy, stress reduction.

PROGNOSIS
Tends to come and go in cycles. Properly treated, it can be controlled.

May affect

20

million in U.S.

assured—because it's now clear that just as a negative mood can make you sick, a positive outlook can make people healthier.

First and most straightforwardly, just as depression can lead people to neglect their bodies and even commit suicide, says Ellen Peters, a psychologist at Ohio State University, "positive moods improve decision making." The reason, according to a 2013 study she coauthored in the journal *Cognition & Emotion*: positive mood changes are believed to trigger the release of the neurotransmitter dopamine into areas of the brain responsible for working memory and judgment. In the study, older adults who were in a good mood during

The chronic stress that millions feel from simply trying to deal with the pressures of modern life can unleash a flood of hormones that are useful in the short term but subtly toxic if they persist.

tests of working memory did an average of 18.6% better than those reporting neutral moods. "There's also evidence from earlier studies that young adults see similar improvements," says Peters.

As a result, people should make better decisions about their health, thus indirectly improving their situation in both the short and long terms. "What seems to matter most," Peters says, "is the effect of a change in mood on decisions that you make afterwards. It's not that you have to be in a positive mood all the time."

There's almost certainly a direct effect as well. The

hormone cortisol, which may be a key to the physical ravages of depression, is also pumped out in excess in the nondepressed when they're under stress. Studies have shown that wounds heal more slowly in people with high stress levels; that stress reduces the immune system's ability to fight off the flu; and that chronic stress can literally shorten lives. Finding ways to reduce emotional stress and increase positive emotion—through practices like meditation, for example—could therefore help the body as well as the mind.

More recently, researchers have found a connection between positive emotion and a key marker of cardiovascular health. Called vagal tone, it's in part a measure of how well your heartbeat returns to normal after an emotionally jarring experience. "It's a very interesting objective measure of health," says Barbara Fredrickson, a psychologist at the University of North Carolina and author of the book *Positivity,* "because it's related to your likelihood of having a heart attack, your ability to regulate your glucose levels, and your immune response."

Recently, Fredrickson and several colleagues published a study in the journal *Psychological Science* that looked at how efforts to improve mood might influence vagal tone. "We taught people a meditation technique that helps them generate positive emotions and increase their attunement to other people," she says. "Those who learned this technique showed improvements in vagal tone." The implication: "Experiencing a steady diet of positive emotions in daily life tunes up our cardiovascular systems in ways that make us physically healthier."

One thing to keep in mind, however, says Fredrickson: "We've found that positive emotions are mild and subtle, while negative emotions kind of hit us like a sledgehammer. They're more intense. So in order to thrive, we need to experience more positive emotions than negative emotions." For some, that seems to come naturally. "People who are more resilient," she says, "tend to experience positive emotions side by side with their negative emotions when they are going through a tough time." So even during a crisis, for example, they'll think, "At least I have the support of my friends and family."

As the study in *Psychological Science* makes clear, some people are simply in the habit of reverting to positive emotion, even in the midst of trouble—but those who aren't can learn to develop that skill. "They can change their emotional habits in lasting ways, in ways that increase their potential for happiness and health. It's not set in stone or DNA," says Fredrickson.

Meditation is one way to do that, but if it seems too hard or New Agey, a 2013 paper in the *British Journal of Health Psychology* suggests that you can improve your mood just by eating better. In a 21-day study of 281 young adults, Tamlin Conner of the University of Otago in New Zealand discovered a strong relationship between a diet high in fruits and vegetables and a posi-

FROM LEFT: YAGI STUDIO/GETTY IMAGES; C.J. BURTON/CORBIS

How Stress Takes Its Toll

Like its more severe cousin depression, ordinary stress is harmful to the body as well as the mind. How it works:

1 A STRESS RESPONSE STARTS IN THE BRAIN

When the brain detects a threat, a number of structures, including the hypothalamus, amygdala and pituitary gland, go on alert. They exchange information with one another and then send signaling hormones and nerve impulses to the rest of the body to prepare for fight or flight.

2 THE BODY UNLEASHES A FLOOD OF HORMONES

Adrenal glands react to the alert by releasing adrenaline, which makes the heart pump faster and the lungs work harder to flood the body with oxygen. The adrenal glands also release extra cortisol and other glucocorticoids, which help the body convert sugars into energy. Nerve cells release norepinephrine, which tenses the muscles and sharpens the senses to prepare for action.

3 THESE HORMONES CAN CAUSE SIGNIFICANT DAMAGE

When the threat passes, epinephrine and norepinephrine levels drop, but if danger comes too often, they can damage the arteries. Chronic low-level stress keeps the glucocorticoids in circulation, leading to a weakened immune system, loss of bone mass, suppression of the reproductive system, and memory problems.

4 THE LESSON, IF YOU WANT TO UNDO THAT DAMAGE

Stay calm. There's nothing wrong with the fight-or-flight response when it's needed. But on the subway or in traffic is not the place. Relax; your health depends on it.

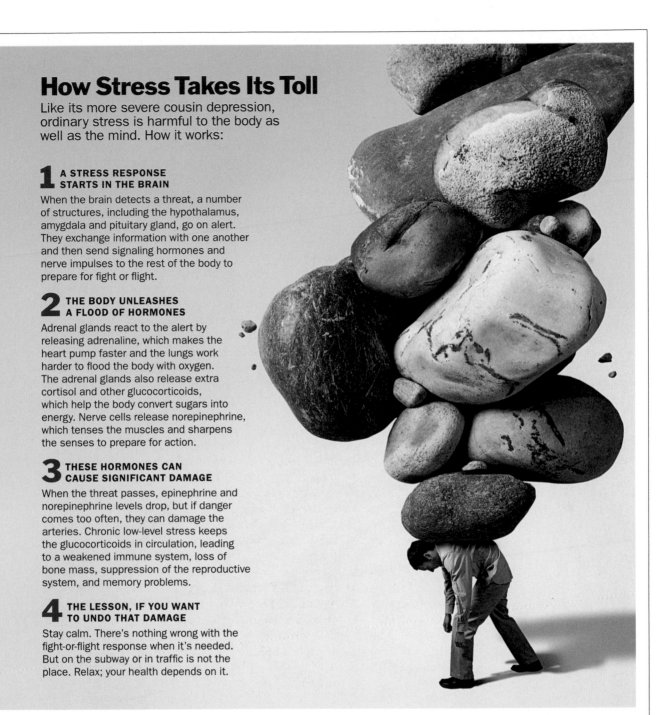

tive mood (see accompanying story "Mood-Boosting Remedies").

"We don't know the precise mechanism of it as yet," says Conner. "It could boost brain serotonin levels. It could be because these foods are high in antioxidants, which can have a calming effect on bodily systems. But it's exciting because it's a whole area of research to be discovered."

Of course, eating healthy foods improves well-being anyway; any extra benefit it might provide through im-provements in mood would just be icing on the cake. But there's nothing wrong with that. As psychiatrists and other doctors are figuring out how to unravel the intimate links and feedbacks between mental and physical health, they're also learning that it doesn't always matter which is cause and which is effect. Whether you can stave off emotional problems by helping the body or stave off physical ills by addressing the mind, the whole person is bound to be better off. —*Reported by David Bjerklie and Jenisha Watts/New York City*

Mood-Boosting Remedies

We've come a long way from Freud's couch. Before the Viennese psychiatrist's day, people were pretty much stuck with whatever mental state they happened to be in. The "talking cure" was a breakthrough in its era, but since then it's been joined by an array of treatments—from drugs to healthy foods—that can help relieve negative moods and boost positive ones.

BY DAVID BJERKLIE

ALTERNATIVE THERAPIES

TODAY'S TREATMENTS More patients today help themselves to over-the-counter aids, from St. John's wort to ginkgo biloba and soybean extracts. But while herbs are natural, they, like prescription drugs, can have side effects, even if they're effective (which is not always clear). The popular supplement DHEA, for example, may be linked to an increased risk of cancer.

ON THE HORIZON It sounds too good to be true, and research is still in its early days, but tantalizing evidence suggests that eating the right foods can contribute to a more positive, optimistic outlook. For example, several surveys by Tamlin Conner, a researcher at the University of Otago in New Zealand, have indicated that people who eat a diet rich in fruits and vegetables—fresh, frozen or even canned—report more positive emotions than those who avoid such foods. (Intriguingly, the produce-heavy diet did not reduce negative emotions such as anxiety, sadness and anger.) The link needs more research, since it's possible that happier people eat more fruits and vegetables. In the meantime, Conner believes it's "probably best for people to eat two-thirds vegetables and one-third fruit," since the effect was stronger for vegetables.

Another key to mood may be a type of omega-3 fatty acid known as DHA, which can help keep electrical signals flowing smoothly in the brain. DHA is found in fatty fish, and research by Dr. Joseph Hibbeln at the National Institutes of Health has shown that depression rates are as much as 65 times as high in countries where people don't eat much fish as in those where fish is a dietary staple. And a number of studies suggest that fish-oil supplements can help combat depression.

ELECTRICAL AND MAGNETIC

TODAY'S TREATMENTS Electroshock therapy, despite its unsavory reputation, is actually quite effective, especially for patients who don't respond to drugs and seniors for whom drug interactions pose problems. The treatment today uses a small current to trigger a mild seizure, which can push a depressed brain out of its rut. Clinicians are also using a similar technique that sends an electrical current through the vagus nerve—a major conduit wiring the heart and intestines—which then delivers it to the brain. Surgery is required to implant the vagus-nerve-stimulation device, so it is generally considered only for people with treatment-resistant depression.

ON THE HORIZON A noninvasive technique called transcranial magnetic stimulation uses an electric coil shaped like a figure eight to create a magnetic field inside the brain's prefrontal cortex, an area that plays a key role in mood regulation. Although the FDA approved the procedure in 2008, much remains unknown about the ideal course of treatment; studies are under way in an effort to answer key questions.

TALK THERAPY

TODAY'S TREATMENTS Most research today is focused on the physiology of depression, yet clinicians find that combining medical and psychological treatments is most effective. Psychoanalysts still try to get patients to probe the unconscious roots of their problems (although Freud's techniques have been adapted and streamlined). Newer techniques like cognitive therapy, in contrast, teach patients to recognize destructive patterns in their lives and develop practical steps for changing bad mental habits. Clinicians who use positive-psychology techniques may recommend something as simple as writing a list of things you're grateful for; this can pull you out of a funk. Helping others is another way to take the focus off your own problems and feel more positive—something self-help gurus and religious leaders have known all along.

ON THE HORIZON Meditation, mindfulness training and biofeedback have long been championed as proven stress relievers. Now proponents believe these techniques may also provide relief to people with depression by lowering levels of cortisol.

DRUGS

TODAY'S TREATMENTS Most antidepressants work by tweaking levels of various neurotransmitters, the chemicals that carry signals from one neuron to another. Prozac, Paxil, Zoloft and other SSRIs slow the absorption of serotonin. Effective antidepressants that act on both serotonin and norepinephrine include Effexor and Remeron. Drugs like Wellbutrin work in a similar way but are believed to aim at the neurotransmitters norepinephrine and dopamine. Older tricyclic antidepressants (such as Elavil and Tofranil) also block the absorption of neurotransmitters, especially norepinephrine, but do so in a different fashion and are more likely to have significant side effects. Another class of drugs, the monoamine oxidase inhibitors (MAOIs) such as Nardil and

Marplan, can be effective but can also produce dangerous side effects. The same is true for low doses of the sex hormone testosterone delivered via transdermal patches. Research shows this treatment can relieve depression, but it can also cause headaches, acne, and swelling of breast tissue.

ON THE HORIZON Researchers are exploring two related molecules, gaba and glutamate, which are responsible for 90% of chemical signaling in the brain. Because they control so much of the brain's activity, the trick is to fine-tune their levels in ways that relieve depression but don't affect other functions. Additional targets of drug development include the stress hormone cortisol, which researchers are trying to regulate with the abortion drug RU-486, as well as compounds called CRF antagonists, which appear to block the action of dynorphins, the evil twins of feel-good endorphins. Also under investigation: a chemical called substance P, involved in pain pathways closely related to depression.

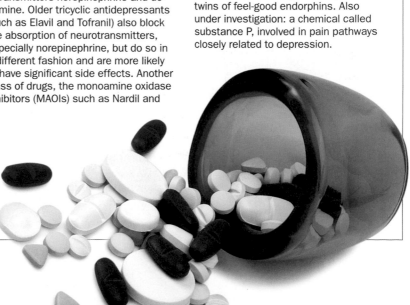

A Primer for Pessimists

Yes, you can learn to see every silver lining, even if you're a born cynic.  BY ALICE PARK

SMOKING AND OBESITY MAY BE the most conspicuous causes of illness in this country, but physical factors don't account for everything. As the previous story showed, your psychology can be just as important to your well-being as exercising and eating right.

In fact, an entire science has grown up around the perils of negative thinking (as well as the power of positive psychology), and findings confirm that a pessimistic outlook not only kindles anxiety, which can put people at risk for chronic mental illnesses like depression, but may also cause early death and set people up for a number of physical ailments, ranging from the common cold to heart disease.

Optimism, meanwhile, is associated with a happier and longer life. Which may be good news for the motivational gurus out there, but what about the rest of us who aren't always so chipper? Are we destined for sickness and failure? Or is it possible to master the principles of positivity the same way we might learn a new hobby?

The answer from the experts seems to be: we can learn to be of good cheer, but it takes effort. Seeing the sunny side doesn't come easily.

Be an "Optimalist" Most people would define optimism as being endlessly happy, with a glass that's perpetually half full. But that's exactly the kind of deluded cheerfulness that practitioners of positive psychology don't recommend. "Healthy optimism means being in touch with reality," says Tal Ben-Shahar, who taught one of Harvard's most popular courses, Positive Psychology, from 2002 to 2008. "It certainly doesn't mean being Pollyanna-ish and thinking everything is wonderful."

Ben-Shahar describes realistic optimists as "optimalists"—not those who believe everything happens for the best, but those who make the best of things that happen. In his own life, he uses three optimalist exercises, which he calls PRP. When he feels down—say, after giving a bad lecture—he grants himself permission (P) to be human. He reminds himself that not every lecture can be a Nobel winner; some will be less effective than others. Next is reconstruction (R). He parses the weak lecture, learning lessons about what works and what doesn't. Finally, there's perspective (P), which involves acknowledging that in the grand scheme of life, one lecture really doesn't matter.

Studies suggest that people who are able to focus on the positive fallout from a negative event can protect themselves from the physical toll of stress and anxiety. In research at the University of California, San Francisco, scientists asked a group of women to give a speech in front of an audience of strangers. On the first day, all the participants said they felt threatened, and showed spikes in fear hormones and the stress hormone cortisol. On subsequent days, however, those women who had reported rebounding from a major life crisis in the past no longer felt the same threat from speaking in public—and did not show a jump in cortisol. They had learned that this negative event, too, would pass. "It's a backdoor to the same positive state because people are able to tolerate the negative," says Elissa Epel, a psychologist involved in the study.

Accept Pain and Sadness Being optimistic doesn't mean shutting out painful emotions. As a clinical psychologist, Martin Seligman, who runs the Positive Psychology Center at the University of Pennsylvania, says he used to feel proud whenever he helped depressed patients rid themselves of sadness, anxiety or anger. "I thought I would get a happy person," he says. "But I never did. What I got was an empty person." That's what prompted him to launch the field of positive psychology, with a groundbreaking address to the American Psychological Association in 1998. Instead of focusing only on lifting misery, he argued, psychologists need to help patients foster good mental health through constructive skills, like Ben-Shahar's PRP. The idea is to teach patients to build their strengths rather than simply ameliorate their weaknesses. "It's not enough to clear away the weeds and underbrush," Seligman says. "If you want roses, you have to plant a rose."

When a loved one dies or you lose your job, you're supposed to feel sad, even depressed. But you can't cocoon yourself in sadness for too long. A study by UCSF researchers of HIV-positive men whose partners had died found that the men who allowed themselves to grieve while also seeking to accept the death were better able to bounce back. Those who focused only on the loss, as opposed to, say, viewing the death as a relief of their partner's suffering, tended to grieve longer, presumably because they couldn't find a way out of their sadness.

Smile in Your Profile Picture If all else fails, try "catching" happiness. We are social beings, and our outlook is influenced by that of our friends and family. Dr. Nicholas Christakis, a professor at Harvard Medical School, documented in a 2008 study just how powerful this network effect is. Compared with glum people, those who were happy were more likely to be surrounded by other happy people. Even the friends of happy people's friends' friends (who might be complete strangers) tended to be happy. Christakis and his colleague James Fowler at the University of California, San Diego, are now studying happiness contagion in Facebook. They noticed that people who smiled in their Facebook profile pictures tended to have friends who smiled. This might simply be peer pressure at work, but Christakis and Fowler are investigating whether it isn't a more infectious phenomenon.

If you still aren't convinced that your doomsaying ways can be changed, consider this: only about 25% of a person's optimism may be hardwired in his genes, according to some studies. That's in contrast to the 40% to 60% heritability of most other personality traits. Science suggests that the greater part of an optimistic outlook can be acquired with the right instruction, a theory borne out in a study of college freshmen by Seligman. Pessimistic students who took a 12-week optimism-training course—which included exercises like writing a letter of gratitude and reading it aloud to someone—were less likely to visit the student health center for illnesses during the next four years than pessimistic peers who weren't tutored in positive thinking.

The thing about being optimistic, though, is that it takes hard work. It's an active process, say psychologists, through which you force yourself to see your life a certain way. Indeed, the leading optimism and happiness experts consider themselves born pessimists. But if they have learned with time and lots of practice to become more hopeful, take heart! So can you.

Just Say Om

Scientists study it. Doctors recommend it.
Millions of Americans—many of whom don't
even own crystals—practice it every day. Why?
Because meditation works.

✳ BY JOEL STEIN

T'S ALMOST LIKE I'M BEING GUILTED INTO IT. LIKE GOING TO THE
gym. Every week I read another study that says if I meditate I'll be
smarter, nicer, happier. In just 20 minutes, I'm told, I can be a better
me. So, fine, I'll stop by the free meditation session that UCLA's Mindful Awareness Research Center throws in the university's Hammer
Museum each week.

I walk into the Billy Wilder Theater and wait as about 200 people
fill the place. And they don't look at all like meditators—no hand-
knitted backpacks or Lululemon pants. In fact, I've never been around
this many old people in Los Angeles when I wasn't at a hospital that
doesn't do cosmetic procedures. I'm not sure if old people like to meditate, like
museums, or just like stuff to do in the middle of the day.

We're waiting for everyone to finish filing in when Gloria Kamler, a meditation teacher at UCLA, sitting on the stage with a cordless microphone, tells us
otherwise. "We're not really waiting, so much as noticing what we're experiencing while we're doing the thing called waiting." This is what I was afraid of.

But once we start the loving-kindness meditation, it's New Age–free. She
tells us to focus on our feet and then our legs, until we settle on our facial muscles
and become very aware of our body. And our breath. And our moment. And I
fall asleep. Seven times. I have little micro-dreams—one about Quentin Tarantino, one about cooking carrots, and one about sex that luckily involves neither
Tarantino nor carrots.

Admittedly, I'm just learning to meditate, and it's hard to be in the mo-

CONTEMPLATIVE FITNESS *Meditation is increasingly used as a means for coping with stress in places where you might not expect it. Clockwise from top: For military cadets at Norwich University in Vermont, training in TM supplies a tool that might help prevent PTSD; students at a San Francisco middle school participate in a "quiet time" session; an inmate at San Quentin prison meditates as part of a yoga class.*

ment when you're also supposed to be remembering all those moments for this article. But for a few seconds I really am present, and it feels like being superalive, like when you're truly enjoying something or totally focused on a task. But, probably due to the seven naps, I leave feeling more tired than invigorated. As Kamler tells me afterward, I tend to be either in monkey mind or asleep, never completely present—I'm always either worrying about what I have to do next or just checked out.

Still, my monkey mind has managed to notice something: since the last time I meditated (once), 10 years ago, meditation has changed. It's no longer religious. It's not even liberal anymore. Just as yoga was severed from its Indian roots, meditation has been released into the marketplace, its Buddhism as obfuscated as the Christianity in opening presents under the tree.

It's even gone corporate. At Google, Chade-Meng Tan—a 42-year-old engineer who was only the 107th employee of the company in 2000—is now in charge of teaching mindfulness to employees. More than 1,000 people at the company have taken his seven-week-long class, Search Inside Yourself, where they've learned to "stop, breathe, notice, reflect and respond." In 2012 Google spun off the class into the nonprofit Search Inside Yourself Leadership Institute, which teaches business leaders how to meditate. Oprah Winfrey paid for her entire staff to learn Transcendental Meditation (or TM), and they now practice daily. Dr. Oz not only did a whole show on the health benefits of meditation but practices himself.

At the beginning of 2013, the U.S. Marines—yes, the Marines—started a pilot program at Camp Pendleton called M-Fit (Mindfulness-Based Mind Fitness Training), designed by former Army captain Elizabeth Stanley, who, like many veterans, used meditation successfully to deal with post-traumatic stress disorder. A group of Marines are learning to meditate to calm themselves in the middle of maneuvers, building on a study the Corps did in 2011 that was so successful, Maj. Gen. Melvin Spiese called meditation "push-ups for the brain."

Meditation has become a tool that helps people focus in a world of constant little distractions known as cell phones. "I don't really enjoy going to the gym, but I do it because the benefits to your cardiovascular health are obvious," says Jay Michaelson, who has taught at Yale and Boston University Law School and whose book *Evolving Dharma* comes out in October 2013. "That's true for meditation, too. It's a physical process that happens in the brain. And that's true whether you think it's for you or not. Just like we have physical fitness in the gym, we are going to have contemplative fitness. And just like we have reading, writing and arithmetic in school, we are also going to have mindfulness and conflict resolution. It's not for weirdos anymore." In fact, meditation is now the stuff of too many TED Talks. There are so many meditation retreats that in 2008 *USA Today* ran an article listing America's top choices. An app called Transcend, which is used with a Bluetooth headset, monitors your brain waves to score your meditation sessions. To end a session, you can download the 99-cent Mindfulness Bell app and set it to ring on your phone.

Ten years ago, the celebrities associated with meditation were impressive but not, by and large, surprising. There was David Lynch (whose David Lynch Foundation for Consciousness-Based Education and World Peace funds research into meditation, along with classes for kids, the homeless, veterans and refugees), Goldie Hawn (whose MindUP program trains schoolteachers to give instruction on meditation to help children with self-control), Paul McCartney (who traveled to India with the other Beatles to meet Maharishi Mahesh Yogi, the founder of TM), Richard Gere (who is friends with the Dalai Lama) and Shirley MacLaine (who is Shirley MacLaine). But today's meditators are less obviously liberal, such as Arnold Schwarzenegger and Clint Eastwood. *Good Morning America* co-host Dan Harris, who has also done reports for *The 700 Club*, has been on weeklong meditation retreats and is writing a book about neuroscience and meditation. South Carolina Republican congressman Mark Sanford (who, as a nonmeditating governor, left his wife to go "hike the Appalachian Trail" with his Argentine mistress) says he meditates now. In 2012, Ohio Democratic congress-

Through the Ages

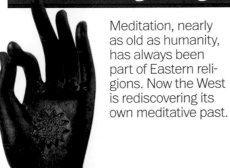

Meditation, nearly as old as humanity, has always been part of Eastern religions. Now the West is rediscovering its own meditative past.

PREHISTORY
Shamanistic Tradition

No one knows precisely when meditation began, but experts think it could well have been practiced by hunter-gatherers many thousands of years ago. Like other mystical practices, it might have been reserved for tribal shamans, who were believed to be in direct touch with the invisible world of spirits.

3000–2000 B.C.
Vedic Tradition

Meditation is described in ancient Hindu texts; it has been a part of Hinduism and its many offshoots ever since.

23

man Tim Ryan put out a book about meditation called *A Mindful Nation* in which he wrote, "If more citizens can reduce stress and increase performance…they will be better equipped to face the challenges of daily life, and to arrive at creative solutions to the challenges facing our nation." The most unlikely practitioner may have emerged on April 21, 2013, when Rupert Murdoch tweeted, "Trying to learn transcendental meditation. Everyone recommends, not that easy to get started, but said to improve everything!"

Advocates of meditation have a lot of studies to prove their case. The first scientific research, in the 1960s and '70s, basically showed that meditators are really, really focused. In India in the 1960s, a researcher named B.K. Anand found that yogis could meditate themselves into trances so deep that they didn't react when hot test tubes were pressed against their arms. Another study showed that master meditators, unlike marksmen, don't flinch at the sound of a gunshot.

In 1967, Dr. Herbert Benson, a professor of medicine at Harvard Medical School, found that meditators used 17% less oxygen, lowered their heart rates by three beats a minute, and increased their theta brain waves—the ones that appear right before sleep—proving that they indeed reached a different state than either normal consciousness or sleeping. In his 1970s bestseller *The Relaxation Response*, Benson, who founded the Mind/Body Medical Institute, argued that meditators counteracted the stress-induced fight-or-flight response and achieved a calmer, happier state. Several years later, Dr. Gregg Jacobs, a professor of psychiatry at Harvard Medical, recorded EEGs of meditators; spikes in their theta waves showed that they essentially deactivated the frontal areas of the brain that receive and process sensory information. They also managed to lower activity in the parietal lobe, a section of the brain located near the top of the head that orients you in space and time. In other words, simply by focusing on the moment, meditators controlled the way their brains receive input.

In 1997, University of Pennsylvania neurologist Andrew Newberg hooked up a group of Buddhist meditators to IVs containing a radioactive dye, which allowed him to prove that the brain doesn't shut off when it meditates but rather blocks information from coming into the parietal lobe. The brain, after being trained through meditation, is choosing not to hear and see information in exchange for focusing on the present state. The University of Wisconsin–Madison's Richard Davidson has used brain imaging to support Benson's long-ago contention: through regular meditation, the brain is reoriented from a stressful fight-or-flight mode to one of acceptance, a shift that increases contentment.

Studies on meditation gained momentum in 2000, when the Dalai Lama met with Western-trained psychologists and neuroscientists in Dharamsala, India. This work was orchestrated by Jon Kabat-Zinn, a molecular biologist who, in 1979, founded the Center for Mindfulness in Medicine, Health Care, and Society at the University of Massachusetts Medical School (see accompanying story "The Art of Mindfulness Meditation"). Over the years, he has helped more than 14,000 people manage their pain without medication by teaching them to focus on what their pain feels like and accept rather than fight it. "These people have cancer, AIDS, chronic pain," he says. "If we think we can do something for them, we're in deep trouble. But if you switch frames of reference and entertain the notion that they may be able to do something for themselves if we put very powerful tools at their disposal, things shift extraordinarily."

Kabat-Zinn has also studied people with psoriasis, a chronic skin disease that is often treated by asking patients to go to a hospital, put goggles on, and stand naked in a hot, loud ultraviolet-light box. Apparently, many people find this stressful. So Kabat-Zinn randomly picked half the patients and taught them to meditate to reduce their stress levels in the light box. In experiments, the meditators' skin cleared up at four times the rate of the nonmeditators. In another study, conducted with the University of Wisconsin's Davidson, Kabat-Zinn gave a group of newly taught meditators and nonmeditators flu shots and measured the antibody levels in their blood. Researchers also mea-

588 B.C.
Buddha
After meditating under a banyan tree, Prince Siddhartha Gautama achieved enlightenment. Buddhism in all its many forms would be the ultimate result.

2ND CENTURY A.D.
Christian Meditation
A group of early Christian monks known as the Desert Fathers retreated from the world to live in simplicity. They used meditation to get closer to God. For more than 1,000 years afterward, meditation would be an increasingly important part of Christian practice.

CIRCA 1000
Cabalistic (Jewish) Meditation
Its practitioners believe that the Jewish mystical tradition of Cabalism is ancient, but it was codified in Europe only about 1,000 years ago. Meditation was one way that Cabalists would try to commune with God.

CELEBRITIES WITH MANTRAS *Clockwise from above: Oprah Winfrey paid for her entire staff to learn TM; Sheryl Crow said she drew serenity from meditation after being diagnosed with breast cancer; Russell Simmons has a book coming out on the practice's scientific benefits; David Lynch's foundation funds research as well as classes for kids, the homeless, veterans and refugees.*

sured their brain activity to see how much the meditators' mental activity shifted from the right brain to the left. Not only did the meditators have more antibodies after they received the shots, but those whose activity shifted the most had even more antibodies. The better your meditation technique, Kabat-Zinn suggests, the healthier your immune system.

Some of his recent studies show that eight weeks of meditation thickens parts of the brain used in learning, memory, executive decision making, and perspective taking, while thinning the amygdala, which deals with fear. And there have been many other studies. One in 2011, by Harvard and MIT researchers, found that the reason meditators can block out pain and distraction is that they are controlling brain waves called alpha rhythms. A 2013 Emory University study miked up new meditators and recorded them at random times during the day; it discovered that they had become a whole lot nicer. Research at U.C. Santa Barbara showed that meditation improved test scores significantly on the reading-comprehension section of the GREs, even though improved nutrition didn't. A 2013 study published in *Psychological Science* suggested that loving-kindness meditation increases the tone of the va-

---

1000
Muslim Meditation

At about the same time that some Jews were embracing mysticism, certain Muslims were doing the same. The Muslim sect known as Sufis (after the plain wool garments, called *suf,* that they wore) incorporated meditation into their rituals of worship.

EARLY 1500S
Martin Luther

He didn't approve of mysticism and preferred a plain reading of Scripture to any kind of incantation. In response to the Reformation he inspired, the Roman Catholic Church suppressed the influence of monks who taught meditation.

1967
Maharishi Mahesh Yogi

Promoting his own brand of meditation, this guru won the Beatles as converts and began a resurgence of meditation in the Western world that still flourishes today.

gus nerve, which, among other things, reduces heart rate and blood pressure. People who have toned vagus nerves have more positive emotions and calm themselves better when they have negative feelings.

One study, in the *International Journal of Geriatric Psychiatry*, suggested that meditation might reduce the risk of dementia. Another, in the American Heart Association's journal, showed that African-American meditators with heart disease were 48% less likely to have a stroke or heart attack than those who just went to a health-education class. And research from Georgia Regents University showed that meditating teens reduced their left ventricular mass, an indicator of future cardiovascular disease. The field is growing so fast that in 2011 there were 397 scholarly papers published on meditation listed by the Institute for Scientific Information, compared with just 28 in 2001. About 6.3 million Americans have been told by doctors to practice meditation or similar activities.

There are so many studies coming out that psychologist and author Daniel Goleman, who is on the board of the Buddhist-oriented Mind & Life Institute, has been running management-training sessions based on some of these findings. His next book is even more directly marketed to the growing meditation-friendly business-book-buying audience—*Focus: The Hidden Driver of Excellence*. In January 2013 the journal *Social Cognitive and Affective Neuroscience* devoted a special issue to "Mindfulness Neuroscience," including studies on how meditation assists smokers in quitting and helps people overcome social-anxiety disorder. Music mogul Russell Simmons, who sits on the board of Lynch's foundation, has a soon-to-be-published 100-page book that focuses on the scientific basis of meditation's benefits. "That research is helpful to the meditator—it's good faith that it works," he says. "It doesn't work until you say it does, to some degree." The proof that meditation works helps people put aside the doubts that get in the way of their meditation practice.

When I meditated 10 years ago, I remember mostly trying not to giggle. But since then, I have had one superpower: by focusing on the present and really feeling my bladder and its fullness, I could get myself to not need to go to the bathroom. I'm not really sure this is a great health benefit, but it does make me believe that all the medical research is true.

Even more telling, I've taught my 4-year-old son that when he gets frustrated and loses control and wants to hit someone, he should take deep breaths, as the Muppets demonstrated on *Sesame Street*, and focus on a spot on the wall. It works for him. And the best part is that I didn't even have to show him any brain-imaging scans to prove it. —*Reporting by David Bjerklie, Alice Park and David Van Biema/New York City, Karen Ann Cullotta/Iowa, and Jeanne McDowell/Los Angeles*

The Art of Mindfulness Meditation

You don't need a guru to find the secret of centeredness. All you have to do is breathe.

BY MAIA SZALAVITZ

THERE ARE COUNTLESS WAYS TO MEDITATE— maybe even an infinite number. But a theme that runs through most practices is mindfulness: a nonjudgmental focus on your immediate sensations, thoughts and feelings, rather than the past or future.

Jon Kabat-Zinn, an MIT-trained biologist, has long been an ambassador of mindfulness, nearly single-handedly bringing the practice into mainstream medicine after being introduced to it himself in 1966. Kabat-Zinn developed the techniques of mindfulness-based stress reduction (MBSR) that have been shown to be beneficial for conditions ranging from hypertension to addiction, and he has personally taught mindfulness to thousands.

But there isn't any "one true way" to get its benefits, he emphasizes. As with exercise or diet, the right routine is the one that you will actually make a regular part of your life. So it's best to begin by choosing a time to practice daily and by trying different approaches.

Getting Started

Sitting quietly on a cushion first thing in the morning and focusing on each inhalation and exhalation for 15 minutes is one way to start, if it works for you. If it doesn't, walking or running (or anything rhythmic and repetitive) is a good alternative. If silence proves difficult, listening to music or chanting a mantra can be another way to invoke a mindful meditative state.

"The real way to start is to be open to experimenting or playing with the possibility of noticing what you're experiencing in this moment, and not [trying] to feel differently," says Kabat-Zinn. "Most people think that to meditate, I should feel a particular special something, and if I don't, then I must be doing something wrong. That is a common but incorrect view."

Getting Bothered

In fact, you can count on one thing: no matter what activity you are doing—sitting, listening or exercising—

your thoughts will tend to wander (or bolt) away from the meditative focus. If they do, the discomfort or desire for change can become your teacher. Thoughts like "I hate this," "I'm a bad meditator" or "This is boring" should simply be acknowledged, but not heeded, as you turn back to your breathing or awareness. "The whole point is to just simply notice the play of the mind and body, and not [take] things personally when they aren't [engaged]," Kabat-Zinn says. "It's good to try 15 minutes, long enough so that you get really bored and antsy and learn how to make room for unpleasant moments."

Indeed, the most important insights of mindfulness often come from being stuck or frustrated by not being able to do it better. Meditation isn't about achieving some goal, whether that goal is meditating "correctly" or becoming calm so that you can perform better in another part of your life. "My definition of healing is coming to terms with things as they are," says Kabat-Zinn, "so that you can do whatever you can to optimize your potential, whether you are living with chronic pain or having a baby."

But don't be surprised if you suddenly feel compelled to check your e-mail or smartphone along the way, he says. "The impulse can come up in the strangest of moments, and you can become aware of how strongly you actually want to distract yourself from the present moment. There might be an important e-mail waiting for you, or something to take you away from this crappy boring moment. Even before smartphones and the Internet, we had many ways to distract ourselves. Now that's compounded by a factor of trillions."

Getting It

This is why mindfulness is especially useful in a stressful and distracting world. "[Meditation] is not about getting anywhere else—it's about being where you are and knowing it," Kabat-Zinn says. So much of our behavior is motivated by urges going on below the surface of our awareness, he adds: give me more or get me out of here. But if you can learn to be mindful, those forces become far less stressful, improving both physical and mental health, no matter what your circumstances.

And you can start anywhere, anytime, just by focusing on your breathing. The best time to begin, of course, is now.

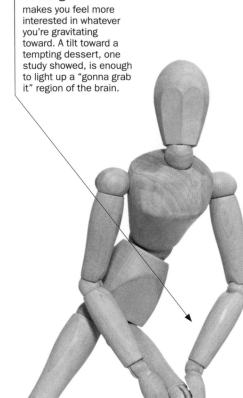

Leaning in
makes you feel more interested in whatever you're gravitating toward. A tilt toward a tempting dessert, one study showed, is enough to light up a "gonna grab it" region of the brain.

Can Posture Change Your Mind?

Surprise: standing tall or hunching over affects more than your appearance.
To soothe an anxious mood (and maybe drop a few pounds), move right this way.

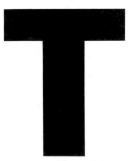 BY REGINA NUZZO

THE MIND-BODY CONNECTION has long been thought of as a rather simple affair. Picture it: a bossy mind orders around its servant body like an aristocrat issuing sniffy requests to a household staff downstairs. "Body, get out of bed. Arm, lift coffee mug. Mouth, we're feeling jolly. Let's have a smile."

But as fans of the PBS drama *Downton Abbey* might have guessed, the real picture is far more interesting. Servants in that early-20th-century English country estate wield a delicious power over their mistresses and masters. Likewise, research increasingly shows, your body downstairs has a surprising amount of influence over your mind up above.

Indeed, something as simple as your posture can affect what your brain thinks, feels and believes. Psychologists call this "embodied cognition," a term that pays homage to deep-seated wisdom hidden in our bodies. Their research is showing that how you move your muscles and where you place your limbs, head and torso all help control your mood, your behavior, and the way you think. Even your facial expressions—the micro-postures of the smallest muscles in your cheeks and around your eyes—can have profound effects. "Normally, being happy causes you to smile. But research shows that making a person put their face into the pose of a smile can lead them to be happy," says psychologist Richard Petty, an embodied-cognition researcher at Ohio State University. "It's as if the happiness circuits go both ways."

With this new view, the mind-body conversation becomes a lot more interesting. "Cheek muscles are now lifting and eyes crinkling up," observes the hapless mind. "That usually happens when we're feeling swell.

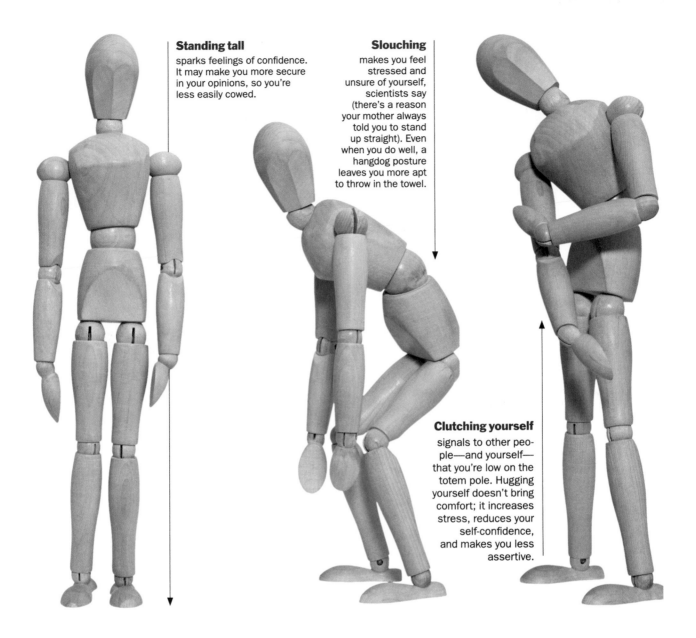

Standing tall
sparks feelings of confidence. It may make you more secure in your opinions, so you're less easily cowed.

Slouching
makes you feel stressed and unsure of yourself, scientists say (there's a reason your mother always told you to stand up straight). Even when you do well, a hangdog posture leaves you more apt to throw in the towel.

Clutching yourself
signals to other people—and yourself—that you're low on the totem pole. Hugging yourself doesn't bring comfort; it increases stress, reduces your self-confidence, and makes you less assertive.

Right, then! Cue the happy feelings!" And thus the body cunningly tricks the mind into feeling certain emotions.

This is more than merely piquant; it could, just possibly, help you win friends, influence people, and lose a little weight. Here's a look at how your body can affect your mind from morning to night.

First stop: coffee shop. You smile brightly. You may start to feel happier even before the caffeine kicks in, because a big smile, even a fake one, activates the feel-good regions of the brain. When people in lab studies activate their smiling muscles, they find cartoons funnier. They're faster to spot happy expressions on others' faces, more easily understand sentences with pleasant messages, and get a bigger kick out of the music they're listening to. One recent study found that smiling even helped people's hearts return more quickly to a relaxed pace after exer-

cise and other stresses—a sign of cardiac health.

What's going on here? Psychologists have a saying: "Neurons that fire together wire together." In other words, we spend so much time smiling when we're happy that eventually the smiling-muscles parts of the brain get hooked up with the feeling-happy parts of the brain. The connection has been reinforced so many times that it doesn't matter whether you have a real reason to feel happy. Fake it and you'll make it.

Next: yoga class. You stretch your arms high above your head. You may start thinking more positive thoughts, says psychologist Daniel Casasanto of the New School for Social Research in New York. "Making simple repetitive movements, either upward or downward, can change what we choose to think about," he says. In his recent study, participants who told stories about their lives while moving marbles from one card-

Don Draper
When you're the alpha dog, you can take up all the room you want. Maybe your legs are planted wide; maybe your elbows are angled out. "Relax," your body is telling your mind. "The world is your oyster."

board box to a higher one offered up more positive memories, such as winning awards. Participants moving marbles from a high box to a lower one recalled more negative ones, like flunking exams.

You scoff? "Up" means "positive" in our brain's mental geography—and yes, it has a noticeable effect. Many studies have shown it: if we see a positive word, we process it faster if it's presented higher in our visual space, while we process "bad-feeling" words faster when they're shown lower. When we decide that a word is positive or negative, our eyes shift up or down accordingly; and the more depressed we feel, the more we focus our attention at the bottom of our visual field.

The mental connection appears to be deeply ingrained, and researchers say it may be because important things in our early lives tended to be situated up high—parents and their hugs, for example, or the exciting objects we could reach when we finally learned to stand up. So when things are good, our spirits soar. When things don't go well, we're down in the dumps and feeling low. With heaven firmly planted up above, is it such a surprise that reaching for it lifts our spirits?

At the office, you adopt a Don Draper pose. It just might unleash a whole new you, says Dana Carney, an assistant professor at the business school of the University of California, Berkeley.

"Power is displayed through large, expansive pos-tures," Carney says. "And if you engage in those postures, you can trick your mind a little bit into feeling like it has power."

You know a power pose when you see it. Think of adman Draper with his arm draped casually on the couch back, or an opera diva with her chest high and arms wide. Their limbs are open and extended; they're taking up every inch of space they appear to believe they deserve. They're relaxed, confident, dominant.

In a recent study, Carney and her colleagues investigated how much these postures matter. They adjusted participants' bodies into a low-power pose (for example, seated with hands folded demurely in the lap) or a high-power one (such as hands clasped behind the head and feet on the desk), and had them hold the position for a minute under the pretext of testing heart-rate equipment. Then the volunteers played a gambling game.

The results were startling. The high-power posers' levels of the stress hormone cortisol dropped, while levels of confidence-inducing testosterone rose. These volunteers even placed more assertive bets than those who'd held submissive postures. How much of a difference can such positioning make? In a similar study, some people held a commanding pose while prepping for a stressful mock job interview, while others were told to sit like a shrinking violet. The psychological residue stuck around long enough that the interviewers saw the power posers as more employable.

At the grocery store, you grab a shopping cart. That might help you resist the candy bars.

Observe the wisdom of a toddler. Yummy sweet potatoes are made to be grabbed and stuffed into the mouth, boring toys to be pushed away and flung to the ground. In fact, for our entire lives, we've pulled toward us the things we like and pushed away those we don't—with the result, researchers say, that the "I want this now" parts of the brain have been wired together with the flexed-arm-muscles parts of the brain. And the regions responsible for "no thanks, not now" feelings are linked with areas responsible for extending your arm muscles.

This may sound like the mutterings of psychologists with too much time on their hands, but it's actually one of the well-studied effects in embodied cognition. Researchers get study participants to unknowingly simulate a pulling motion, usually by having them press up on a tabletop from below (and a pushing motion by pressing down from above). This work has shown that when people are engaged in this pulling-toward-the-body posture, they drink more sweet beverages, desire advertised products more, opt for smaller rewards now rather than bigger rewards later, and eat more chocolate cookies.

In one recent study, researchers from Erasmus University in the Netherlands assigned shoppers to either carry a basket tucked next to their body with their arm flexed or push a cart with their arms extended. Result: the basket carriers were much more likely than the cart pushers to pick Twix and Mars bars over apples and oranges. The researchers suggest that this effect might help explain why, outside the lab, people find slot machines so seductive. After all, they require you to pull a lever toward your body, thus tapping into brain circuits that choose vice over virtue.

Embodied cognition is a booming field in psychology, and researchers are hard at work figuring out all the ways that our bodies can lead our minds, that motion can lead to emotion. There's just one small catch: the postures might work only if you don't realize you're doing them. "It is not yet clear if we can use these techniques ourselves, or if the techniques work because people are unaware of them," says Ohio State's Petty. When you know how a magic trick is done, it's hard to un-know what you know and believe the magician just sawed his assistant in half. The same goes for bodies and minds.

Still, it's easy to dream of a future in which we know just the right posture to pull ourselves out of every bad mood. A future where expert frame-of-mind coaches prescribe personalized body contortions to start the morning right, and office up-and-comers are encouraged to stretch out and put their feet up on conference tables. And where diet-friendly refrigerators are designed so that you have to push the freezer door to get at the ice cream—ice cream that, thanks to the pushing, you will no longer even want.

Kate Middleton
True joy is signaled by a smile that engages the squinty crow's-feet muscles around the eyes. Called a Duchenne smile, it's hard (but not impossible) to fake.

Pat Riley
What posture screams "persistent to the point of pigheaded"? Arms crossed, chin jutting, shoulders spread wide—the famed basketball coach had it down. It's the stance of determination.

31

Staying Healthy

THE DIET THAT MIGHT SAVE YOUR LIFE
THE FIRES WITHIN
MONITORED ME ■ THE ANGELINA EFFECT
WHY WE DON'T ACT OUR AGE

THANKS TO ADVANCES IN DRUGS AND VACCINES, not to mention public sanitation, epidemics no longer regularly roar through large swaths of the developed world, snuffing lives like a fast-moving fire. Now researchers are trying to figure out how to quench slower-burning dangers like heart disease, cancer, and the gradual decay of old age. And the more closely scientists look, the simpler some of the solutions seem. The right diet may be more powerful than we've ever realized, and common-sense health habits may play a big role in damping down the quiet, whole-body conflagration—inflammation—that appears to underlie many of the biggest killers. As people become obsessive observers of their own bodies, ceaselessly monitoring their behavior and tweaking it for better results, how much time can they buy? Let's explore.

The Diet That Might Save Your Life

The news on the Mediterranean diet makes it seem as if it can do everything but shine your shoes and pay your rent. Is it really as good as it sounds?

✳ BY ALEXANDRA SIFFERLIN

F YOU WANDER INTO THE STORAGE BASE-ment of Amáli, a Mediterranean restaurant on New York City's Upper East Side, you'll be hard-pressed to find anything canned. Instead, freezers and shelves are stuffed with a wide range of fresh vegetables, a little cheese, plenty of wine, and lots and lots of olive oil. "We go through about 200 liters of olive oil a week," says executive chef Nilton Borges Jr. "I marinate with it, I poach with it, I love the flavor." Borges prepares his dishes in the Mediterranean style—simply, maximizing flavor without all the extra "stuff." There are no cream-based sauces or heavy marinades; vegetable dishes are not drenched in cheese. Borges's broccoli dish needs just a little lemon juice, olive oil, salt, pepper and herbs, and it's ready for the table.

Mediterranean-style meals are typically undemanding when it comes to preparation, but don't let the simplicity fool you. The diet—characterized by olive oil, fruits and vegetables, legumes and grains; a moderate amount of nuts, dairy (mostly in the form of cheese), wine and protein (primarily in the form of fish); and not much saturated fat or sweets—adds up to something special.

Researchers who look at such things have long viewed the Mediterranean diet as a healthy one, but in the past five years or so, its effects have started to seem nothing short of stunning. Eating this way can protect diabetics from the worst consequences of their disease, studies have shown. The diet can reduce the risk for Alzheimer's disease and keep waistlines trim. The most impressive finding of all? It may actually lengthen your life. And it tastes good, too.

The Mediterranean diet first gained notice in the 1950s, when University of Minnesota physiologist Ancel Keys started the Seven Countries Study to investigate the relationship between diet and heart disease in various regions of the world. Among its results, Keys's study found indications that the Mediterranean diet significantly lowered the risk of coronary heart disease.

Soon scientists started noticing evidence of other health benefits. For one thing, there seemed to be quite a few unusually healthy elderly people in countries that followed the diet. Rates of a number of cancers seemed oddly low as well. Studies piled up for a few decades until, in 1993, researchers from around the world came together in Boston for a conference convened by the World Health Organization, the Harvard School of Public Health and a nonprofit organization called Oldways Preservation and Exchange Trust. Billed as an opportunity to review the research, the conference was also a way for scientists to publicize the diet's benefits,

FEAST FOR THE SENSES
*With its healthy and delicious
ingredients, the Mediterranean diet
offers a remarkable array of benefits.*

A WAY OF LIFE *No one food supplies the diet's health-giving powers. What does it is the combination of nutrients you get from regularly eating this way—as well as the things you avoid, such as a lot of meat and dairy.*

and resulted in the first Mediterranean Diet Pyramid. Suddenly, people from Manhattan to Mendocino began to be interested.

Nowadays, shoppers take it as a matter of course that they can find olive oil at their grocery store in a multitude of grades (plain, virgin and extra-virgin). They can buy hummus that tastes not just like chickpeas and cumin but like red pepper, pesto or jalapeño. As for yogurt, the Greek variety has become a go-to item for foodies and brown-baggers alike. At the same time, the pace of research has accelerated. A 2013 study sums things up—scientists analyzed nearly 200 traditional

Greek foods and found that the evidence supported 1,024 nutritional claims for these Mediterranean diet staples. Here are some of the most impressive findings for the diet as a whole:

It Keeps Your Heart Healthy

The largest study to date on the Mediterranean diet and heart disease, published in April 2013, showed that this way of eating lowers the risk of cardiovascular disease and stroke by as much as 30%. The study involved 7,447 people in Spain who were overweight, smoked or had other heart-disease risk factors, such as diabetes. With his colleagues, Dr. Ramón Estruch of the Hospital Clinic of Barcelona put the volunteers on either a low-fat diet or one of two versions of the Mediterranean diet, one high in extra-virgin olive oil and the other heavy on nuts.

Compared with those eating a low-fat diet, people on a high-oil version of the Mediterranean diet had a

30%

lower risk of suffering a heart attack or stroke or dying of heart disease.

---

switching from a meat-heavy Western diet to Mediterranean fare can help keep waistlines in check. The investigators, at Boston Children's Hospital, were looking at 21 overweight or obese volunteers who had already lost weight, to see what kind of dietary regimen would make it easier to keep the pounds off (often the hardest part of weight loss). The study participants took turns on three different diets: low-fat, low-carb and what's called a low-glycemic-index diet. That last one is made up of foods that are absorbed more slowly into the body, keeping blood sugar stable and insulin levels normalized. It bore a striking resemblance to the Mediterranean diet, consisting largely of whole grains, lean protein like fish and beans, fruits and vegetables, and healthy fats from olive oil and nuts.

When dieters ate from a low-carb menu, they burned the most calories—a full 325 more each day than when on the low-fat diet. That weight-loss boost didn't come for free, however. The low-carb diet caused increases in the stress hormone cortisol and in C-reactive protein, a risk factor for heart disease. But people on the Mediterranean-style diet got the best of both worlds. They showed no heart-unhealthy changes yet burned 150 calories more each day than when they ate the low-fat diet. That's roughly the equivalent of an hour of moderate activity, without breaking a sweat.

To James Mallios, co-owner of Amáli, findings like these come as no surprise. "I battled with weight my whole life," he says. "But when I would go on vacation with my family in Greece, I wouldn't watch what I ate at all, but I would come back and have lost weight."

Compared with those eating the low-fat diet, people on the high-oil version of the Mediterranean diet had a 30% lower risk of suffering a heart attack or stroke or of dying of heart disease during the study. Those consuming the Mediterranean diet with more nuts showed a 28% lower risk. "A reduction of 30% in the relative risk [of having] a cardiovascular event is really a reduction similar to the effects of some drugs, but without adverse effects," says Estruch. "We presumed the Mediterranean diet would have a powerful effect, but not to such an extent."

It Helps You Manage Your Weight

Obesity is one of the biggest health problems facing Americans, indirectly causing an estimated 300,000 premature deaths every year. More than a third of U.S. adults qualify as obese, but a 2012 study suggests that

It Sharpens Your Brain

Since the beginning of 2012, we've seen many of the largest-ever studies of the Mediterranean diet. In addition to the heart-health research, a different massive study looked at the diet's benefits to the brain. Researchers at the University of Alabama at Birmingham and the University of Athens in Greece tracked more than 17,000 African-American and Caucasian men and women for four years and found that those who closely followed a Mediterranean diet were 13% less likely to develop memory and

SIMPLE AND SUSTAINING *Mediterranean-style meals are typically undemanding to prepare, maximizing flavor without all the rich extras.*

thinking problems than other participants. An earlier study found that people who stuck to such a diet were less likely than others to show evidence on an MRI of having had a small stroke. In fact, not eating a Mediterranean diet boosted the risk of these memory-threatening brain changes about as much as having high blood pressure.

It Can Lengthen Your Life

Back in the 1960s, Dr. Antonia Trichopoulou of the University of Athens spent hours videotaping old Greek village women as they cooked, asking them what they ate and how they made it. In the years since, she's become one of the most active researchers in this field. In 2003, her study of more than 22,000 Greek older adults found that those who most closely adhered to the Mediterranean diet had a survival advantage: they were 25% less likely to die during the follow-up period than others she tracked. But the benefits are not reserved solely for those residing in Mediterranean regions. In a 2005 study of over 74,000 elderly men and women from much of Europe, Trichopoulou showed that people in these countries who ate a Mediterranean-

style diet lived as much as 14% longer than people with different eating habits. "Longevity has been proven by several studies now," Trichopoulou says. "I am so happy to see the recognition this diet is gaining as a valuable diet for the world."

What's Behind the Benefits

Most researchers agree that no single food is responsible for the health-giving powers of the Mediterranean diet; instead, the combination of nutrients you get from regularly eating this way is what makes it so special—along with the nutrients you avoid. Following a Mediterranean diet means curbing your consumption of meat and dairy, and that's good for your heart. Eating plenty of monounsaturated fats, found in olive oil and nuts, increases HDL ("good") cholesterol. Filling your plate with fruits and vegetables guarantees lots of flavonoids, antioxidants that reduce inflammation. That's a good thing, since most chronic diseases are related to oxidation and inflammation, says Estruch. Wine, in moderation, is thought to enhance the effects of these nutrients. And all the antioxidants combined are thought to protect the brain against oxidative stress.

BOUNTY OF GOODNESS
Greek salad is a classic of the region, perfect for warm weather.

What's more, because Mediterranean fare is full of low-glycemic-index foods, it keeps your blood sugar steady, which helps control appetite. The fact that it's relatively high in fat may also help you manage your weight, because fat keeps you satiated.

But eating native foods is probably not the only thing that accounts for the health of people in the region. Traditionally, life in this part of the world has included lots of activity; for people living a rural, agrarian life, there's no need to go to the gym for exercise. Food is also more than mere sustenance in the Mediterranean culture. It's a focal point for socializing. Sitting down to a family meal is common practice, and research has shown that families who eat together stay well together, with trimmer waistlines and a lower risk of depression.

Still, there's no denying that the Mediterranean way is a very good one to follow. And although there aren't an abundance of olive groves in the U.S., and most people here don't live in fishing villages, there are easy ways to incorporate Mediterranean habits into the average American diet—and do it economically.

Those who closely followed a Mediterranean diet were

13%

less likely to develop memory and thinking problems.

Eating more fruits and vegetables is critical, but they can be frozen or canned; they should fill half your plate. Cooking with olive oil instead of butter reduces your intake of saturated fat while boosting your consumption of antioxidants. Instead of eating red meat, try fish and legumes. Experiment with healthier options for treats, too. You'd be surprised how quickly you start to appreciate the sweetness of an apple or banana and honey for dessert.

Remember that making time for meals with family and friends is also inherently Mediterranean, and that the shared experience brings both pleasure and health perks. As a bonus, making these changes could mean that you end up consuming fewer calories without even noticing.

At Amáli, communal tables encourage customers to socialize with one another over leisurely meals. The restaurant tends to be full of conversation and laughter. "The Mediterranean style is more than just a diet," says Mallios of the atmosphere he seeks to foster. "In this case, the word 'diet' is not about deprivation. Following the Mediterranean diet is a lifestyle."

Mediterranean Shopping List

1) FISH *offers lots of protein without saturated fat, and is high in vitamin D and omega-3 fatty acids—all of which protect your heart.* **2) SPICES** *are full of antioxidants and add flavor without sodium, which may help keep blood pressure down.* **3)** *Packed with fiber and antioxidants,* **VEGETABLES** *reduce the risk of stroke, heart disease and (possibly) cancer and help control weight.* **4)** *The Mediterranean diet is relatively low in dairy, but does feature* **FETA CHEESE,** *a source of calcium and vitamin D, which aid bone health.*

5) *The vibrant colors of* **FRUIT** *mean it's high in vitamins and antioxidants that encourage heart, memory and urinary-tract health.* **6) WHOLE GRAINS** *lower cholesterol, blood sugar and blood pressure.* **7) BEANS** *are a great protein source; their potassium and magnesium help keep blood pressure in check.* **8)** *With more than 200 micronutrients,* **OLIVE OIL** *lowers cholesterol and helps control blood sugar.* **9)** *A compound in* **RED WINE** *called resveratrol has been linked to a lower risk of heart disease and diabetes and a longer life.*

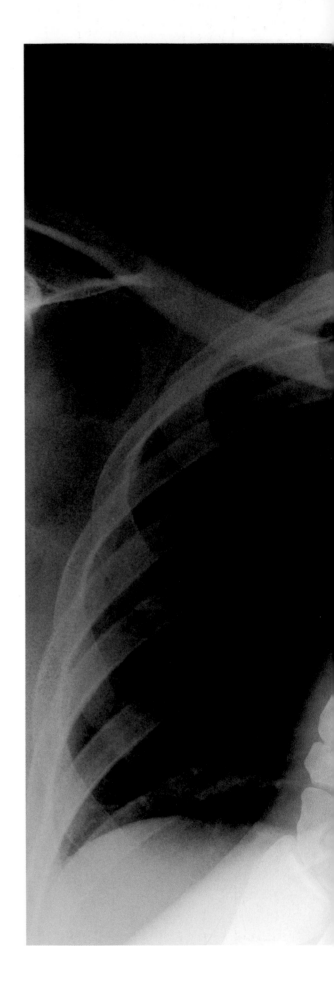

The Fires Within

How inflammation may be to blame for some of the biggest modern killers—and the key to curing or preventing it.

 BY CHRISTINE GORMAN, ALICE PARK AND KRISTINA DELL

WHAT DOES A STUBBED TOE or a splinter in a finger have to do with your risk of developing Alzheimer's disease, suffering a heart attack or succumbing to colon cancer? More than you might think. As scientists delve deeper into the causes of illness, they are starting to see links to an age-old immunological defense mechanism called inflammation—the same process that turns the tissue around a splinter red and causes swelling in an injured toe. And there is growing evidence that this understanding could radically change doctors' concepts of what makes us sick and lead to new ways to keep us well.

Most of the time, inflammation is a lifesaver that enables our bodies to fend off disease-causing bacteria, viruses and parasites. The instant any of these potentially deadly microbes slips into the body, inflammation marshals an attack that lays waste to both invader and any tissue it may have infected. Then, just as quickly, the process subsides and healing begins.

Every once in a while, however, the feverish production doesn't shut down on cue, and inflammation becomes chronic. When that occurs, the body turns on itself like an ornery child who can't resist picking a scab, with aftereffects that

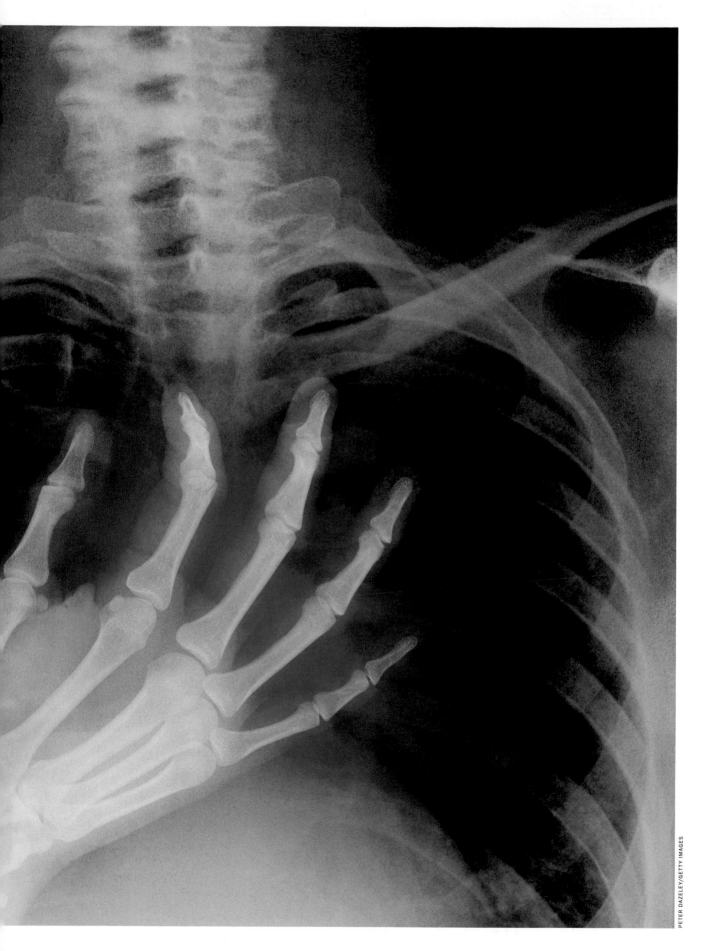

43

seem to underlie a wide variety of diseases.

In recent years, chronic inflammation has become one of the hottest areas of medical research. Hardly a week goes by without the publication of a study uncovering a new way that chronic inflammation does harm. It destabilizes cholesterol deposits in the coronary arteries, potentially leading to heart attack and stroke. It chews up nerve cells in the brains of Alzheimer's victims. It may even foster the proliferation of abnormal cells, egging them on toward becoming cancerous. In other words, chronic inflammation may be the engine that drives many of the most feared diseases of middle and old age.

The concept is intriguing because it suggests a new and possibly much simpler way of warding off illness. Instead of different treatments for, say, heart disease, Alzheimer's and colon cancer, there might be inflammation-reducing strategies that could be used to prevent all three.

This view is changing the way some scientists do research. "Virtually our entire R&D effort is [now] focused on inflammation and cancer," says Dr. Robert Tepper, a partner at Third Rock Ventures, a firm that is focused on launching health-care companies with entrepreneurs. Across the U.S., cardiologists, rheumatologists, oncologists, allergists and neurologists are discovering that they're all looking at the same thing. Just a few years ago, "nobody was interested in this stuff," says Dr. Paul Ridker, a cardiologist at Brigham and Women's Hospital in Boston who has done groundbreaking work in the area. Now the field even has its own peer-reviewed publication, the *Journal of Inflammation Research*, which publishes papers on such topics as the molecular mechanisms of the condition and novel anti-inflammatory drugs.

To understand better what all the excitement is about, it helps to know a little about what happens when the body is subjected to trauma or injury. As soon as that splinter slices into your finger, sentinel cells stationed throughout the body alert the immune system to the presence of any bacteria that have come along for the ride. Some of those cells, called mast cells, release a chemical called histamine that makes nearby capillaries leaky. This allows small amounts of plasma to pour out, slowing down invading bacteria, and prepares the way for other faraway immune defenders to enter the fray. Meanwhile, another group of sentinels, called macrophages, begin a counterattack and release more chemicals, called cytokines, which signal for reinforcements. Soon, wave after wave of immune cells flood the site, destroying pathogens and damaged tissue alike.

Doctors call this generalized response to practically any kind of attack innate immunity. Even animals as primitive as starfish defend themselves this way. But higher organisms have also developed a second, precision-guided defense system that helps intensify the innate response, while also creating antibodies custom-made to target specific kinds of bacteria or viruses. This so-called learned immunity is what enables drug companies to develop vaccines against diseases like smallpox and flu. Working in tandem, the innate and learned immunological defenses fight pitched battles until all the invading germs are annihilated. Then the inflammatory process recedes and healing begins.

Unless, for one reason or another, the inflammatory process persists and becomes chronic. In that case, the final effects vary, depending in part on where in the body the runaway reaction takes hold.

Is Your Heart on Fire?

Not long ago, most doctors thought of heart attacks as primarily a plumbing problem. Over the years, fatty deposits slowly built up on the insides of major coronary arteries until they cut off the supply of blood to a vital part of the heart. A molecule called LDL, the so-called bad cholesterol, provided the raw material for these deposits. Clearly, anyone with high LDL levels was at greater risk of developing heart disease.

There's just one problem with that explanation: half of all heart attacks occur in people with normal cholesterol levels. Not only that, but as imaging techniques improved, doctors found that the most dangerous plaques weren't necessarily all that large. Something that hadn't yet been identified was causing the deposits to burst, triggering major clots that cut off the coronary blood supply. In the 1990s, Ridker became convinced that some sort of inflammatory reaction was responsible for the bursting plaques, and he set about trying to prove it.

To test his hunch, Ridker needed a blood test that could serve as a marker for chronic inflammation. He settled on C-reactive protein (CRP), a molecule produced by the liver in response to an inflammatory signal. During an acute illness, like a severe bacterial infection, levels of CRP shoot from less than 10 mg/L to 1,000 mg/L or more. But Ridker was more interested in the low levels of CRP—less than 10 mg/L—that he found in otherwise healthy people, and that indicated only a slightly elevated inflammation level. In fact, the difference between normal and elevated is so small it must be measured by a specially designed assay called a high-sensitivity CRP test.

By 1997, Ridker and his colleagues had shown that healthy middle-aged men with the highest CRP levels were three times as likely to suffer a heart attack in the next six years as those with the lowest levels. In 2008, he and his colleagues followed up with a study showing that patients with high CRP levels but normal cholesterol levels who took statin drugs halved their risk of having a heart "event" compared with similar patients who did not take the medications.

Inflammation's harm to the heart seems to kick in when levels of CRP reach 3 mg/L or higher; those amounts can triple the risk of heart disease. By contrast, folks with extremely low levels of CRP, less than 0.5 mg/L, rarely have heart attacks.

"This is not about replacing cholesterol as a risk

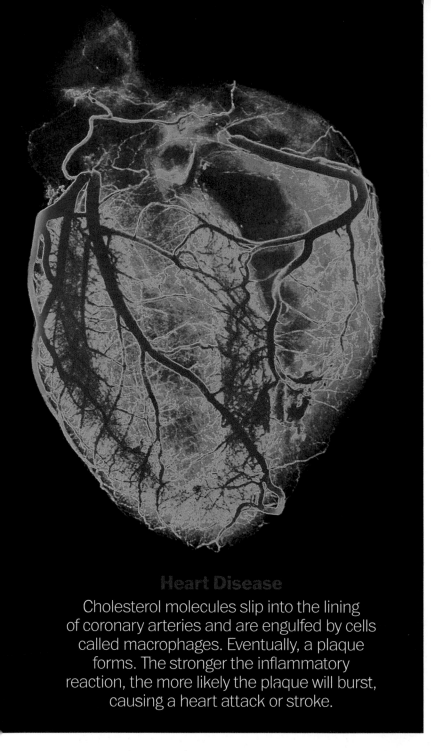

Heart Disease

Cholesterol molecules slip into the lining of coronary arteries and are engulfed by cells called macrophages. Eventually, a plaque forms. The stronger the inflammatory reaction, the more likely the plaque will burst, causing a heart attack or stroke.

A New View of Diabetes

Before Dr. Frederick Banting and his colleagues at the University of Toronto isolated insulin in the 1920s, doctors were desperate. They tried to treat diabetes with morphine, heroin—and high doses of salicylates, a group of aspirin-like compounds. Sure enough, the salicylate approach reduced sugar levels, but at a high price: side effects included a constant ringing in the ears, headaches and dizziness. Today's treatments for diabetes are much safer and generally work by replacing insulin, boosting its production, or helping the body make more efficient use of the hormone. But researchers have been reexamining the salicylate approach for new clues about how diabetes develops.

What they have discovered is a complex interplay among inflammation, insulin and fat—either in the diet or in large folds under the skin. (Fat cells behave a lot like immune cells, spewing out inflammatory cytokines, particularly as you gain weight.) Where inflammation fits into this scenario remains unclear, but the case for a central role is getting stronger. Dr. Steven Shoelson, a senior investigator at the Joslin Diabetes Center in Boston, has bred a strain of mice whose fat cells are supercharged inflammation factories. The mice become less efficient at using insulin and go on to develop diabetes. "We can reproduce the whole syndrome just by inciting inflammation," Shoelson says.

That finding suggests that a well-timed intervention in the inflammatory process might reverse some of the effects of diabetes. Indeed, some of the drugs that are already used to treat the disorder, like metformin, may work because they also dampen the inflammation response.

factor," Ridker says. "Cholesterol deposits, high blood pressure, smoking—all contribute to the development of underlying plaques. What inflammation seems to contribute is the propensity of those plaques to rupture and cause a heart attack."

Cardiologists are not ready to recommend that the general population be screened for inflammation levels. But the American Heart Association says CRP could be helpful in guiding treatment strategies for those with a moderately elevated risk of cardiovascular disease. At the very least, a high CRP level might tip the balance in favor of more aggressive therapy with treatments—such as aspirin and statins—that are already known to work.

Cancer: The Wound That Never Heals

Back in the 1860s, renowned pathologist Rudolf Virchow speculated that cancerous tumors arise at the site of chronic inflammation. A century later, oncologists paid more attention to the role that various genetic mutations play in promoting abnormal growths that eventually become malignant. Now researchers are exploring the possibility that mutation and inflammation are mutually reinforcing processes that, left unchecked, can transform normal cells into tumors.

How might that happen? Some of the most potent weapons produced by macrophages and other inflammatory cells are the so-called oxygen free radicals. These molecules devastate just about anything that crosses their path, particularly DNA. A glancing blow that damages but doesn't destroy a cell could lead to a genetic mutation that allows it to keep on growing and dividing. To the immune system, the abnormal growth looks very much like a wound that needs to be fixed, says Lisa Coussens, chair of the cell and developmental biology department at Oregon Health & Science University. "When immune cells get called in, they bring growth factors and a whole slew of proteins that call other inflammatory cells," she explains. "Those things come in and go 'heal, heal, heal.' But instead of healing, you're 'feeding, feeding, feeding,'" which may allow the abnormal growth to transform into cancer.

Scientists are exploring the role of an enzyme called cyclo-oxygenase-2 (COX-2) in the development of colon cancer. COX-2 is yet another protein produced by the body during inflammation. Researchers have shown that folks who take daily doses of aspirin—which is known to block COX-2—are less likely to develop precancerous growths called polyps. The problem with aspirin, however, is that it can also cause internal bleeding. Celebrex, another COX-2 inhibitor that is less likely than aspirin to cause bleeding, also reduces the number of polyps in the large intestine. Unfortunately, studies have shown that COX-2 inhibitors are associated with an increased risk of cardiac events, and the Food and Drug Administration has asked makers of the drugs to include a boxed warning alerting doctors and patients to the risks.

Aspirin for Alzheimer's Disease?

When doctors treating Alzheimer's patients took a closer look at who seemed to be succumbing to the disease, they uncovered a tantalizing clue: those who were already taking anti-inflammatory drugs for arthritis or heart disease tended to develop the disorder later than those who weren't. Perhaps the immune system mistakenly saw the plaques and tangles that build up in the brains of Alzheimer's patients as damaged tissue that needed to be cleared out. If so, the ensuing inflammatory reaction seemed to be doing more harm than good. Blocking it with anti-inflammatories might limit, or at least delay, damage to cognitive functions.

The most likely culprits are the glial cells, whose job is to nourish and communicate with the neurons. Researchers have discovered that glial cells can produce inflammatory cytokines that call additional immune cells into action. "The glial cells are trying to return the brain to a normal state," explains Linda Van Eldik, an anatomy and neurobiology professor at the University of Kentucky. "But in neurodegenerative diseases like Alzheimer's, the process seems to be out of control. You get chronic glial activation, which results in an inflammatory state."

Preliminary research suggests that low-dose aspirin and fish-oil capsules—both of which are known to reduce inflammatory cytokines—seem to reduce a person's risk of Alzheimer's disease. Unfortunately, most of these preventive measures need to be started well before any neurological problems develop. "It's very hard to improve people who already have [dementia]," says Dr. Ernst Schaefer, a senior scientist at Tufts University. "But it may be possible to stabilize people and to prevent disease."

Quenching the Flames

Everywhere they turn, doctors are finding evidence that inflammation plays a larger role in chronic diseases than they thought.

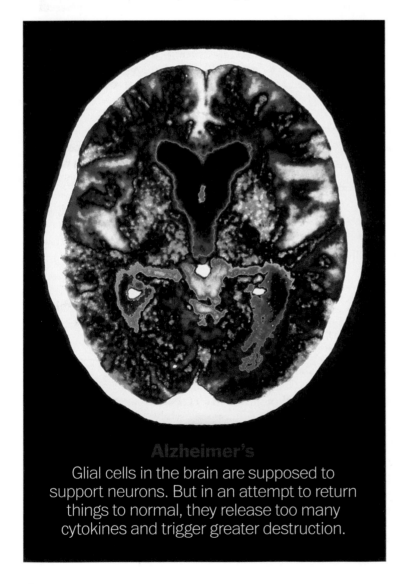

Alzheimer's
Glial cells in the brain are supposed to support neurons. But in an attempt to return things to normal, they release too many cytokines and trigger greater destruction.

What You Can Do
How to keep the damaging effects of inflammation to a minimum.

BY ALICE PARK

DIET

Unsaturated fats: It's not clear yet which dietary fats fight inflammation best, but it makes sense to avoid the saturated fats in red meat and dairy products and stick with fish and vegetable oils.

Fruits and vegetables: The richer in color the better, since colorful plants tend to have the most antioxidants—good for mopping up free radicals produced during inflammation.

DRUGS

Aspirin: A well-known inflammation fighter, aspirin can cool reactions raging in heart arteries and the colon. Similar agents are also showing promise in controlling inflammation in the brains of Alzheimer's patients.

Statins: Not only do they lower cholesterol, but statins also drive down levels of CRP and other inflammatory proteins.

Beta blockers and ACE inhibitors: Doctors are investigating whether blood-pressure medications control hypertension in part by lowering levels of certain inflammatory factors that constrict the blood vessels.

EXERCISE

It should be no surprise that being active is good for you, but inflammation may finally explain why. Fat cells are efficient factories for producing key inflammatory elements, and burning calories shrinks those cells. With fewer elements around, inflammation is less likely to flare up or get into the slow burn that contributes to heart disease, hypertension and diabetes.

ORAL HYGIENE

Keeping your mouth clean by flossing and brushing regularly can reduce the risk of gum disease, a source of chronic inflammation.

But that doesn't necessarily mean they know what to do about it.

That may soon change. Researchers are looking beyond aspirin and other multipurpose medications to experimental drugs that block inflammation more precisely. But there is a sense that much more basic research into the nature of inflammation needs to be done before scientists understand how best to limit the damage in chronic diseases.

In the meantime, there are things we can all do to dampen our inflammatory fires. Some of the advice may sound terribly familiar, but we have fresh reasons to follow through. Losing weight induces those fat cells—remember them?—to produce fewer cytokines. So does regular exercise, 30 minutes a day most days of the week. Flossing your teeth combats gum disease, another source of chronic inflammation. Fruits, vegetables and fish are full of substances that disable free radicals.

So if you want to stop inflammation, get off that couch and head to the greenmarket. And try not to stub your toe on the way. —*With reporting by Dan Cray/ Los Angeles*

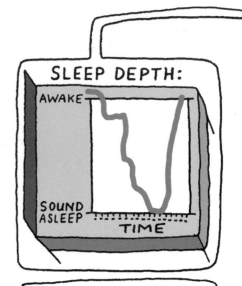

SLEEP DEPTH:

BREAKFAST:

	CHOCOLATE FROSTED DONUT	10 oz COFFEE WITH WHOLE MILK
CALS	270	25
FAT	15g	1g
SODIUM	340mg	20mg
SUGAR	13g	1g
PROTEIN	3g	1g
CARBS	31g	2g
FIBER	1g	0g

MOOD @ 9:15 AM:

Monitored Me

If you want to change it, measure it. The latest trend in health requires quantifying everything about yourself.

✳ BY BRYAN WALSH

HE UNEXAMINED LIFE IS NOT WORTH LIVING." So said Socrates, and I'm trying to live up to the philosopher's credo—in a 21st-century way. In my pocket I carry a Fitbit One, a pen-cap-size device that tracks my steps and calories burned in the course of the day. To make sure I don't exceed those calories out with calories in, I track my diet with the iPhone app MyFitnessPal. The app Sleep Cycle uses my phone's accelerometer to track my sleep patterns over the course of the night. I'm training for a half-marathon, so I track my runs with a GPS watch, which allows me to see myself very slowly getting faster. I track my emotional state with another app that pings me throughout the day, asking me to note my mood and my activity. (Perhaps unsurprisingly, I'm usually at my most unhappy during my daily subway commute.)

That kind of quotidian detail probably isn't what Socrates meant, but more and more of us are engaging in digital self-examination. Nearly 10 million self-tracking devices were purchased in 2011, according to ABI Research, and the number is almost certainly growing as the technology improves. The movement even has a name—the Quantified Self—and its geekiest adherents go far beyond what I do. They carry digital cameras around their necks that capture a constant stream of visual memories, and wear heart monitors and blood-pressure sensors, turning their bodies into walking hospital rooms. But they—we—share a hope: that through collecting ever more information about our bodies and our behavior, we can find a better route to self-improvement. And maybe even accomplish

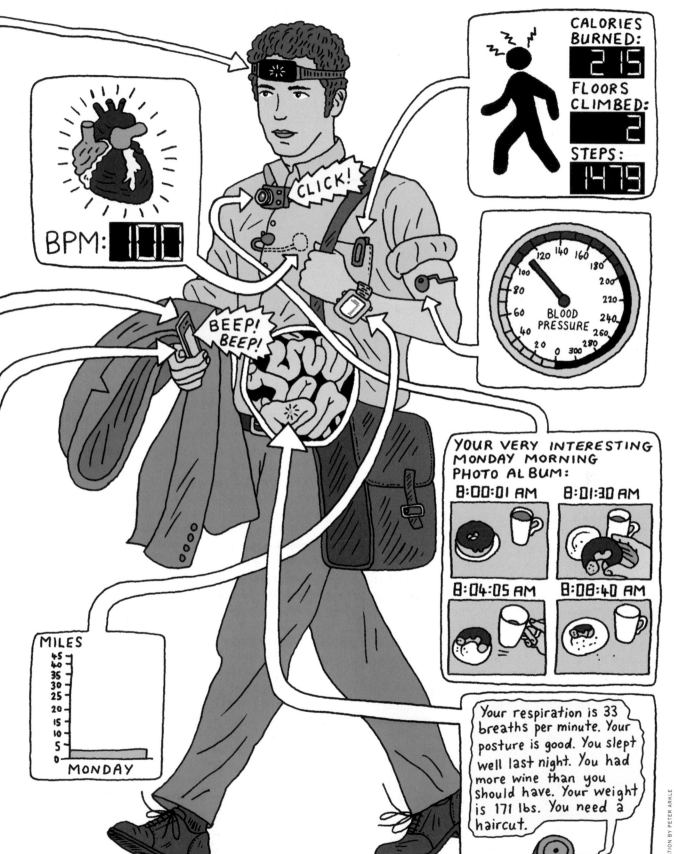

something grander. "This is also for self-knowledge," said Gary Wolf, a journalist and founder of the Quantified Self movement, in a 2010 speech at the TED conference, which marked the moment when this campaign really began to take off. "If we want to act more effectively in the world, we have to get to know ourselves better." To know thyself, you need to know thy data.

Of course, logging personal information is hardly new. Benjamin Franklin kept a meticulous chart book noting his progress on 13 virtues, and dieters in programs like Weight Watchers have long counted calories. As anyone who's ever tried to keep a regular journal knows, though, recording it all on paper requires a commitment few of us can keep up for long. Digital self-tracking devices—often connected to the Internet or our smartphones—take the effort out of recording and compiling. You get better, more regular data, and it's harder for you to smudge it to make yourself look good. And once this information becomes easy to gather and share with doctors, it could revolutionize how we treat

> The point of self-tracking isn't just to accumulate data, though sometimes it can feel that way. Self-trackers want to convert that data into better behavior, better habits—in short, better health.

everything from weight problems to cancer. "Data is the currency of health," says Lisa Kennedy, chief marketing officer of GE's Healthyimagination division, which has gotten involved in self-tracking. "It will help us understand diseases, find new treatments, and even save lives."

The point of self-tracking isn't just to accumulate data, though sometimes it can feel that way. Self-trackers want to convert that data into better behavior, better habits—in short, better health. So analyzing it is at least as important as collecting it. Take my Sleep Mobile app. It's always been a mystery to me why I wake up refreshed on some mornings and exhausted on others, when I've spent roughly the same amount of time in bed. With Sleep Mobile, I've come to understand that certain factors—stress, alcohol, caffeine—can influence how much of that time in bed is actually spent in deep, restorative sleep. I place my iPhone face-down next to my pillow. The phone's accelerometer detects whether I'm moving and, through that, roughly whether I'm awake or in a light or deep sleep. The next morning, the app displays a graphic summary of my night and even gives me a sleep-quality score. That feedback has induced me to give up caffeine in the afternoon and iPad time before bed—with the result that I find myself sleeping better most nights and feeling happier most mornings.

Sleep Mobile is a pretty basic app, and it's probably less accurate than sophisticated sleep-tracking devices sold by companies like Zeo. But Zeo's sleep tracker is ex-

pensive—nearly $200—and requires the sleeper to wear an electronic headband in bed. I have too little disposable income to buy a Zeo, and too much self-respect to use one. As long as self-tracking devices are still relatively expensive and a bit bulky, like the Zeo, the number of Americans who go the full Quantified Self is likely to remain small. That will change as manufacturers invent devices that are more easily worn—imagine sensors sewn directly into clothing—or even designed to be implanted. Proteus Digital Health is making "ingestible event markers" (IEMs) that are the size of a grain of sand. Once swallowed, an IEM sends a high-frequency electric current directly through the body's conductive tissue, to be read on a smartphone. Heart rate, respiration, posture, sleeping patterns and other data can all be reported and recorded, and the user doesn't have to carry anything other than a phone. This may be the sort of thing that Sonny Vu, CEO of Misfit Wearables, meant when he said at the 2013 SXSW conference that the industry shouldn't compromise on wearability, or fashion, for that matter. "Quantified Self will really work when we can wear devices that don't make us look like Tron," he said.

When that happens, self-tracking could produce a health revolution. Right now, doctors have to wait for us to feel bad enough to bring our bodies into the shop; until we do, they're in the dark. Sophisticated self-tracking devices could change that. The Pittsburgh-based company BodyMedia makes sensors that collect 5,000 data points per minute, including measurements of heat flux (the rate at which heat dissipates from the body), motion and skin temperature, all of which can be converted into highly accurate data on calories burned and metabolism. That information could make it a lot easier for someone who is, for example, trying to manage a weight problem—especially if it could be automatically uploaded to a doctor's office. No more lying about how much you exercise or snack. It might be a bit embarrassing, but given how many conditions today require daily maintenance by the patient—think diabetes, which afflicts more than 25 million Americans—that kind of monitoring could mean the difference between life and death.

Self-tracking could help reduce health costs by automating basic health management and diagnosis. And it could help us live longer by pulling a signal out of the noise of day-to-day data to detect illness earlier. But as someone who began practicing self-quantifying for this story and has since become all but addicted to it, I can say there's a personal side to this movement as well. So much of our health today feels out of our hands, the province of medical professionals. Self-quantifying has allowed me to take control of my health, to track and tweak my habits, to make myself a better person. "The quantified self is going to see the rise of the DIY lab," says GE's Kennedy. Today I feel like a test group of one—but I'm in charge of the experiment, and I benefit from the results. Not a bad trade-off for looking a little geeky.

Track Stars

New devices give you even fewer
excuses not to stay healthy.

BY HARRY MCCRACKEN

GADGETS HAVE A LONG AND DIS-
tinguished history of helping
human beings be more sedentary
(exhibit A: the TV remote con-
trol). But the latest generation of battery-
powered wristbands and clip-on devices
aims to reverse that trend, encouraging
healthy activity by tracking how many steps
you've taken and calories you've burned.
The Larklife bracelet will even tell you to
get off your duff: "Noticed you've been sitting for a while," its iPhone app chided me.

Far more engaging than any puny old
pedometer, these accelerometer-equipped
gizmos work with smartphone apps and
cloud-based services to let you monitor your
progress and—unlike some earlier contend-
ers like the Bodybugg—don't charge an
ongoing service fee. Two of them, Larklife
and Jawbone's Up band, are designed to be
worn around the clock and can wake you
during the lightest part of your sleep cycle
so you'll feel more refreshed than if you slept
a bit longer.

Aesthetically, the sinewy Up is the least
obviously nerdy device of the ones I tried.
(It looks identical to the model the company
yanked off the market in 2011 because of
problems with water resistance.) And its
feature-packed app offers the most stuff for
folks who want to get healthy or stay that
way. One nice, if labor-intensive, option: its
app can turn your iPhone into a bar-code
scanner to help you keep tabs on what
you're eating.

The Nike+ FuelBand doesn't track diet
or sleep habits, but it's the gadget I can most
easily imagine sticking with for the long
haul. That's because it has an LED display
that let me check how I was doing without
having to futz with its app. Simply seeing
that little screen, knowing it was quietly
paying attention, made me more likely to
hike up stairs, run errands on foot, and
otherwise prove that I take my health as
seriously as it does.

JAWBONE'S UP, $130

What's hot:
Better-looking than most such
gadgets, with an ambitious
iPhone app that can scan bar
codes on food labels to help
track nutrition info.

What's not:
You can't sync it
wirelessly, and chances
are fairly good that
you'll misplace its tiny
recharging cable.

NIKE'S NIKE+ FUELBAND, $150

What's hot:
The LED display shows how
close you are to meeting your
daily activity goal, and the
app lets you compete against
friends. Also, it's easy to
charge the band from your PC's
USB port, no cable needed.

What's not:
Unlike some rival devices,
it doesn't monitor sleep or
eating habits.

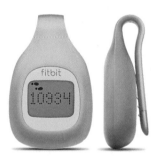

FITBIT'S ZIP, $60

What's hot:
It has a built-in screen and a
battery that lasts for months.
The app (for the iPhone and
Samsung Galaxy) lets you set a
weight goal and log your meals.

What's not:
It's a clip-on, not a wristband.
(Of the fitness gadgets I tested,
this is the only one I kept forget-
ting at home on my dresser.)

LARK'S LARKLIFE, $150

What's hot:
This gadget syncs with an
iPhone wirelessly and automatically. And it comes with
swappable day and night
bands so you can wear one
while charging the other.

What's not:
It's not water-resistant,
and the app is relatively
rudimentary.

STAR POWER
*The example set by Jolie
(here in April 2013) will likely
lead more women to be tested.*

The Angelina Effect

Her preventive mastectomy
raises important issues
about genes, health and risk.

✳ BY JEFFREY KLUGER AND ALICE PARK

THERE'S A CHILLY ARITHMETIC to the way we all get sick. At the end of any year, a fixed and knowable number of us will have developed heart disease, and another number won't have. There will be a different entry in the ledger for cancer, another for lung disease, another for Parkinson's or dementia or HIV. The people who study those mortal metrics—the actuaries, the epidemiologists—don't give too much thought to the individuals behind the numbers, and the truth is, they can't. It's no good sentimentalizing math, not if you want to get anything useful out of it.

But sometimes it's impossible not to: sometimes the person who is sick has a very recognizable face. So it was in 1985, when Rock Hudson, Hollywood heartthrob of an earlier era, died of complications from AIDS, and a country that thought it could fence off a disease suddenly realized we were in this together. So it was in 1995, when Christopher Reeve, a man best known for playing a character utterly immune to injury, was thrown from a horse and suddenly could do nothing at all without help—and with that, the spinal-injury community had a point man a lot more powerful than Superman.

And so it was again in May 2013, when Angelina Jolie, the most beautiful woman in the world by many

Jolie lost her mother, Marcheline Bertrand, to ovarian cancer in 2007. In the U.S., 36% of women who, like Jolie, carry the BRCA gene mutation have preventive double mastectomies.

people's lights, stepped forward and announced in an op-ed in the *New York Times* that she had undergone a double mastectomy, an operation she decided to have after learning she carried a genetic mutation that in her case increased the odds of developing breast cancer to a terrible 87% and ovarian cancer to 50%. She decided to get tested because her mother had died of ovarian cancer at age 56. (Less than two weeks after Jolie's announcement, her mother's sister succumbed to breast cancer.) She herself has no current signs of either disease.

Jolie explained her treatment decision with simple clarity: "Once I knew that this was my reality, I decided to be proactive and minimize the risk as much as I could." She explained it with an eye toward the 12% of all women who will one day develop breast cancer and the 100% who worry about it: "I hope that other women can benefit from my experience. Cancer is still a word that strikes fear in people's hearts." And she explained it in a way that went straight to what many were thinking when a woman whose very name signals beauty and whose profession depends on it made such a dramatic choice. "On a personal note," she wrote, "I do not feel any less of a woman. I feel empowered that I made a strong choice that in no way diminishes my femininity." Jolie, like many other women who have undergone mastectomy, has had successful reconstructive surgery. But as with those other women—especially the ones who were not yet sick—it took a lot of courage to get to that point in the first place. "It's such an emotional and personal decision," says Sarah Hawley, associate professor of general medicine at the University of Michigan, "particularly because it's the woman's choice."

Jolie, according to most experts who have weighed in publicly, made a smart choice for her case. "It's one of the truly unique situations where most medical professionals would say if a woman chose to have both breasts removed, it's a pretty reasonable thing to do," says Dr. Eric Winer of the Dana-Farber Cancer Institute in Boston. Exceedingly reasonable, judging by the numbers. Jolie's doctor estimates that her cancer risk fell from its 87% high to just 5%.

But the seeming straightforwardness of her case masks a much murkier reality, one that involves science, policy and probabilities, not to mention Americans'—really, everyone's—tendency to observe what the famous do and then conclude we should do the same. When Katie Couric underwent a televised colonoscopy in 2000, demand for the procedure jumped, a phenomenon that was promptly dubbed "the Couric effect." In that case, many lives may have been saved by the raised awareness. This trendsetting power is exponentially greater in the case of Jolie, a megawatt star. She gave birth to a

daughter in 2008 and named her Vivienne, and in 2009 that name cracked the top 1,000 in popularity for newborn girls for the first time in the U.S. since 1930. It is now trading at a high of No. 322. Something similar happened with the names of her children Maddox and Shiloh. It's one thing when you model your fashions after Jolie's; it's another thing to model your kids.

If form holds, the 250,000 women each year who undergo the same genetic testing Jolie had will be joined by thousands more. But the mutation that was detected in her, in what's known as the BRCA1 gene, is present in only 0.24% of the population and accounts for no more than 10% of all cases of breast cancer. Still, form does appear to be holding. "I think we will see an increase over the next months for sure in genetic testing for breast cancer," says Rebecca Nagy, a genetic counselor at Ohio State University's medical center and president of the National Society of Genetic Counselors. "What's important to know is that it's not appropriate to test everybody."

And therein lies the problem. In the case of the BRCA genes, a mutation can mean a significant increase in risk. But Otis Brawley, chief medical officer for the American Cancer Society, recalls a woman with no family history of breast cancer who insisted on getting screened for BRCA anyway. The test revealed a mutation of "unknown significance." She nonetheless had a double mastectomy—and the mutation that her test detected has since been shown not to be associated with a higher risk of breast cancer. A growing number of women who discover cancer in one breast are electing to have both breasts removed protectively, even without evidence

Of her family she says, "I can tell my children that they don't need to fear they will lose me to breast cancer… they know that I love them and will do anything to be with them as long as I can."

that they are at genetic risk of having the disease spread. That kind of overreaction, Brawley argues, reflects "the pinking of America," the high-profile campaign to raise awareness about the risk of breast cancer: "We have over-emphasized and scared people too much."

More challenging for doctors trying to guide patients through their choices is the fact that many cancer-screening tests, especially nongenetic ones, do not yield clear treatment options. For some common tests, what looks like trouble may be nothing—or little—of the kind. Thyroid-cancer diagnoses are triple what they were in 1975 simply because doctors are checking more closely for any trace of the disease, but the mortality rate in all those years has not budged. In 2012, experts began recommending that men stop getting routine PSA screening tests for prostate cancer, or at least get them less frequently, since the elevated enzyme levels that may indicate the presence of the disease can also be a result of inflammation, infection, or simply riding a bicycle. Even when the cancer is real, in many cases it grows so slowly that, as doctors say, the patient would have died with it, not of it. For every 1,000 men in the 55-to-70 age group who undergo annual PSA testing over the course of 10 years, a single life will be saved. Meantime, up to 200 will undergo a biopsy, and up to 100 will have their prostate removed unnecessarily.

The problem, of course, is that when it comes to life and death, we don't think about statistical significance and sample groups of 1,000; we think of sample groups of one—and we're the only member. The U.S. may indeed be home to some of the world's best medical technology,

but the final decisions about what to do with all that wondrous know-how still rest with the least rational, most capricious part of the whole system: us. Jolie, to all appearances, made a sober and well-thought-through choice. But every patient is different, and the gravitational pull of a superstar role model has a way of distorting what needs to be a highly individual decision.

The Blueprint of Disease The BRCA1 gene, which sits at the center of Jolie's case, was discovered in 1994 by a team of researchers at the University of Utah and elsewhere. It produces a protein that helps stabilize DNA. Many proteins do this kind of housekeeping chore in many tissues, but BRCA1 is expressed at higher levels in breast tissue, and when it can't do its job, it leaves a lot of room for some of the most defective and destructive cells of all: cancer cells.

Just a year after BRCA1 was discovered, researchers unearthed the BRCA2 gene, which produces a different protein but does basically the same work. Both belong to the group known as tumor-suppressor genes, and certain defects in each can increase the risk of other cancers too, including ovarian and pancreatic and, in men, testicular, prostate, and the rarer male form of breast cancer.

Family history also plays a role in breast cancer, though in ways that aren't entirely clear. According to the American Cancer Society, a woman's risk of developing breast cancer doubles if she has one first-degree relative—a sister, mother or daughter—with the disease; for those with two such relatives, it triples. Much of this data collection was done before widespread testing for

BRCA, and thus all cases of the disease—among women who had the BRCA mutation and those who didn't—were considered to be in one undifferentiated group.

Once BRCA screening became available, you'd have thought the variations in breast-cancer risk would have immediately gotten clearer, but the opposite turned out to be true. A 2007 study found that women whose close relatives tested positive for the BRCA mutation were at up to five times the average risk of developing breast cancer themselves, even if they tested negative for the mutation. A 2011 study, however, overturned that research, finding flaws in the methodology. When those errors were corrected and a different sample group was studied, women without the BRCA mutation who are relatives to women who do have it were at no significantly greater risk of breast cancer than the general population. That points to the dangers of reading too much into even a peer-reviewed study, much less the case of just one woman.

Overlearning the Lesson Whether a double mastectomy was in fact necessary in Jolie's case will almost certainly never be known. If she remains cancer-free, it will be easy to infer a cause and effect, but that doesn't mean that even a woman whose case seems identical to hers would not have other viable options. The anti-cancer drug tamoxifen, taken prophylactically, may cut cancer risk by 40% to 50%, according to recent studies. That, coupled with regular MRI screenings to detect the earliest signs of a tumor, may bring the danger down even further. In the case of an actual malignancy, partial breast removal may suffice as well. "The survival rate for women with early-stage breast cancer who get unilateral mastectomy or lumpectomy with radiation is equivalent," says the University of Michigan's Hawley.

Women who do opt for prophylactic surgery may choose an oophorectomy—removal of the ovaries, something Jolie considered before deciding to start with breast removal, since that surgery is less complex and her risk of breast cancer was higher than her risk of ovarian cancer. But ovarian cancer is also deadlier, mostly because it's harder to detect, and once it does show itself it's often too far along to be curable. There is also good evidence that removal of the ovaries, which produce the estrogen that helps fuel some cancers, can reduce the risk of breast cancer. A 2002 study led by Dr. Kenneth Offit, chief of clinical genetics at Memorial Sloan-Kettering Cancer Center and the discoverer of the most common BRCA2 gene mutation, found that 3% of women undergoing oophorectomy developed breast cancer after about two years, compared with 11% of women who did not have the surgery. Removal of the breasts does seem to reduce the incidence of ovarian cancer—by 89%, according to one 2010 study—but the mechanism is unclear, and the findings are mixed.

"We don't have good screening strategies for ovarian cancer, so it makes sense to try to be aggressive in pre-venting the development of the disease," says Dr. Isabelle Bedrosian, associate professor of surgical oncology at MD Anderson Cancer Center.

Up and down the disease spectrum, holes in our detection screens make these kinds of judgment calls necessary, even when, as with BRCA, we have culpable genes in hand. A study published in 2012 in *The American Journal of Human Genetics* found that with Type 2 diabetes and rheumatoid arthritis, which have been tied to 31 different gene variants each, lifestyle factors such as smoking and obesity—and, in the case of rheumatoid arthritis, a history of breast-feeding—did as good a job of predicting the diseases as reading the genetic tea leaves. Other research has turned up similar results for heart disease. A new genetic-screening test of the tissue in prostate-cancer tumors can help distinguish between aggressive and less severe cases, which may clarify treatment options once the disease has taken hold, but that still leaves the value of PSA testing open to question. Genes are a factor, sometimes a critical one, in diagnosing and treating disease, but they're by no means the only one.

The Map of Health Just how proactive any one patient will be in the face of any one set of risk factors can have as much to do with geography as genealogy. In the U.S., about 36% of BRCA-mutation-positive women opt for preventive double mastectomies; in different parts of Europe, the numbers go in entirely different directions. "If you go to Paris and the Institut Curie and you have a BRCA mutation," says Offit, "the chances of having preventive breast surgery are almost zero. Whereas in Northern Europe, the rate is close to 100%."

The fact that Americans are so proactive in seeking preventive treatment doesn't mean patients can demand any test they might have heard about and think they need. A BRCA-screening test, which can cost up to $3,000, requires a referral, and that's something most doctors—sensibly—don't give out to just anyone who asks. False positives and unclear results can lead to the kind of premature surgery Brawley cites. Even if a doctor does agree that a test is warranted, there's currently no guarantee that a woman's insurance company will pay for it.

Insurers instead place women on a coverage continuum, relying in part on guidelines established by the National Comprehensive Cancer Network (NCCN), an advisory panel made up of some 30 physicians, genetic counselors and other experts. Likeliest to have a BRCA test covered: women with early-onset breast cancer and a close family member who is BRCA-mutation-positive. In the middle are women who do not have cancer but have a family member who tested positive. At the low end are those who do not have cancer and have no close relatives known to be BRCA-mutation-positive.

But the test is only the first part of the process. Also critical is genetic counseling, which isn't free either.

Then there's the cost of any surgery that follows and the reconstruction that may come after that. Depending on individual policies, every one of those stages could represent a separate insurance tollgate, and that leads some experts to wonder if it's even fair to start a patient down that road if she doesn't have the financial means to follow it all the way. "You almost wonder, should I get someone tested if they can't use that information?" says Dr. Mary Daly, the NCCN's chair of the genetic/familial high-risk-assessment panel for breast and ovarian cancers. "It's kind of like doing free mammograms when you don't have a surgeon."

Even after full implementation of Obamacare, the language that guides insurance companies will be vague. Women will be covered for BRCA testing and genetic counseling "if appropriate" and when their "family history is associated with an increased risk for deleterious mutations." The policy vagueness regarding testing and treatment may reflect persistent conflict over health-care priorities. But the scientific uncertainty is unavoidable. Genetic screening is, if not in its infancy, barely out of childhood. And the battle against all disease—especially cancer—is one we've been waging for millennia. The wisdom gained in the lab needs to be matched by the wisdom of both caregivers and patients, and that requires hard thinking and reasoned discussion.

Jolie's role in all this adds one more important dimension. She has long been a symbol of the feminine ideal, which in its shorthand sense has meant feminine beauty. Her body has been a key dimension of her fame; now it may be an even bigger dimension of her influence. The loveliest and most resonant passages in her op-ed piece came during the brief description of her breast reconstruction: "The results can be beautiful," she reassured us, adding that her children can see the small scars but that other than that, "everything else is just Mommy." With that, one of the most stunning women in the world redefined beauty. That made us all a little smarter.

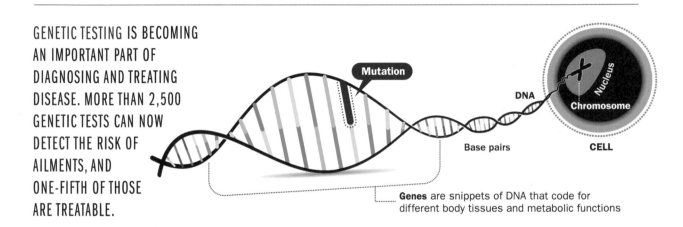

GENETIC TESTING IS BECOMING AN IMPORTANT PART OF DIAGNOSING AND TREATING DISEASE. MORE THAN 2,500 GENETIC TESTS CAN NOW DETECT THE RISK OF AILMENTS, AND ONE-FIFTH OF THOSE ARE TREATABLE.

Mutation

DNA

Nucleus

Chromosome

Base pairs

CELL

Genes are snippets of DNA that code for different body tissues and metabolic functions

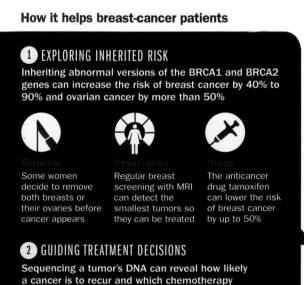

How it helps breast-cancer patients

① EXPLORING INHERITED RISK
Inheriting abnormal versions of the BRCA1 and BRCA2 genes can increase the risk of breast cancer by 40% to 90% and ovarian cancer by more than 50%

Removal
Some women decide to remove both breasts or their ovaries before cancer appears

Observation
Regular breast screening with MRI can detect the smallest tumors so they can be treated

Drugs
The anticancer drug tamoxifen can lower the risk of breast cancer by up to 50%

② GUIDING TREATMENT DECISIONS
Sequencing a tumor's DNA can reveal how likely a cancer is to recur and which chemotherapy drugs will be most effective

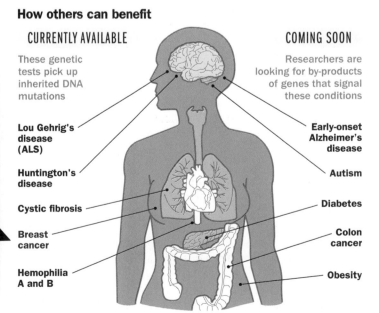

How others can benefit

CURRENTLY AVAILABLE
These genetic tests pick up inherited DNA mutations

Lou Gehrig's disease (ALS)

Huntington's disease

Cystic fibrosis

Breast cancer

Hemophilia A and B

COMING SOON
Researchers are looking for by-products of genes that signal these conditions

Early-onset Alzheimer's disease

Autism

Diabetes

Colon cancer

Obesity

IMMERSION THERAPY
*Synchronized swimmers in
Sun City, Arizona—a community
for "active adults aged 55 and
better"—before a performance.*

Why We Don't Act Our Age

There's no getting around death (at least so far), but aging? To a generation weaned on scientific miracles, that's negotiable.

✳ BY CATHERINE MAYER

F ONE PLACE ON EARTH HAS VANQUISHED nature and stopped the clocks, it is Las Vegas. Built on land without water or any reliable resource apart from the blazing sun, the resort entombs visitors in the permanent dusk of hotel casinos. I had come to this confected city to find out if the Cenegenics Medical Institute, "the world's largest age-management practice," could subvert the laws of human biology with similar ease. First I had to locate Cenegenics, and though you might think it would be easy to spot a building described by its tenants as "quite a lot like the White House," the cabdriver took more than a few passes before we were able to pick out the right White House from the rows of White Houses that have sprouted in the Nevada desert.

That's the Vegas paradox: despite the mind-boggling range of architectural styles and eras represented, there's a remarkable uniformity to it all. The residents are similarly homogeneous. Perma-tanned and toned, many of them sport a uniface common to both genders and across the income range, from bellhops to casino owners. It looks neither young nor old. It is ageless. It is amortal.

Amortality—the term I coined for the burgeoning trend of living agelessly—is a product of the world many of us now inhabit. Childhood, adolescence, young adulthood, middle age, retirement, golden years, decline: each milestone used to be benchmarked against a series of culturally determined ideals. But as our life spans have lengthened—across the developed world, we are now living 30 years longer than we were at the beginning of the 20th century—the meaning of age has become elusive. Children dress like louche adults. Their parents slouch around in hoodies and sneakers.

The rules of age-appropriate behavior that used to be reliably drummed into us by parents and teachers no longer hold sway. But we haven't lost faith; we've just transferred it, to scientists and celebrities. "I think you should just keep going while you can, doing what you like," Mick Jagger observed at 66, ignoring his pronouncement in 1975 that he'd rather be dead than singing "Satisfaction" at—or presumably long after—45.

Doing what you like might include adopting children at 49 and 50, like Madonna; becoming a first-time dad at 62, like Elton John; or marrying a woman 60 years younger than yourself, like Hugh Hefner. The defining characteristic of amortals is that they live the same way from their late teens right up until death. They rarely ask themselves if their behavior is age-appropriate, because

> The defining characteristic of amortals is that they live the same way from their late teens right up until death. The concept of age-appropriate behavior has little meaning for them.

that concept has little meaning for them. Amortals assume all options are always open.

Upturning conventions of aging isn't necessarily a bad thing. By 2050, 27% of the U.S. population will be 60 or older. The amortal impulse to stay active could help ease the anticipated labor shortage and curb swelling health-care costs. But this way of thinking is not without its risks. Amortals have a dangerous habit of trusting that science will be able to deliver them from the consequences of aging or, at a minimum, allow them to select the timing and manner of their passing.

Gains in longevity have been achieved by eliminating or neutralizing many threats to our lives, but the main threat—aging—has proved more resistant to intervention. In 1961 a microbiologist named Leonard Hayflick made a depressing discovery. He found that most human cells are able to divide only a limited number of times, so that even if we get through life without contracting a single disease, we'll die when enough of our cells cease dividing. Although our life expectancy continues to increase, the "Hayflick limit" would appear to doom us to a maximum of around 120 years.

But that doesn't stop amortals from aspiring to spend as long in their bodies as possible. There's a thriving health-care sector promising to help us do so, and Las Vegas is one of its hubs.

The Youth Industry

Eighty percent of Cenegenics patients are men who go there in search of Life. That's Dr. Jeffry Life (his real name), the star of a press campaign in which the physician wears snug shorts and an undershirt with scooped armholes, a style popular in New York City's Greenwich Village. His face is avuncular. His body is that of Mr. Universe in his prime.

Cenegenics describes its program as a "unique and balanced combination of nutrition, exercise and hormone optimization," which sounds good for just about anyone, so I submit myself as a guinea pig. My day at the clinic involves blood work, scans and tests and would usually cost $3,400. (I accepted a free consultation.) At 49, my physical condition turns out to be good but not "optimal," the Cenegenics buzzword. My consultant, Dr. Jeffrey Leake, another paragon of muscularity, tells me that the transition to "optimal" would entail a fierce program of exercise and losing 10 to 15 pounds. Although my body-mass index was logged at 19, toward the lower end of normal, one of the scans detected visceral fat, a hazard for heart disease, diabetes, hypertension and bad cholesterol.

Leake also proposes starting me on two steroid hormones—DHEA and testosterone. "We will monitor for possible side effects of androgen therapy, which are acne, oiliness of skin or deepening of the voice," he e-mails later. The sex hormone estradiol, like testosterone, could improve my bone density, he adds. "And if applicable—only after a comprehensive evaluation reveals an adult-onset growth-hormone deficiency—we may consider supplementing with a third, growth hormone."

Underpinning these recommendations is the notion that hormone supplementation can return us to our "natural" state of youthful vibrancy, as if aging itself were against nature. It's an enticing sales pitch that helps persuade patients to sign up for further consultations and supplies of supplements that typically add up to around $1,000 per month.

Clients often arrive on Cenegenics's doorstep in bad shape; weaned off junk food and coaxed into the gym, converts are likely to show clear benefits. But there is another reason Cenegenics patients often achieve such dramatic results: testosterone and other steroids promote muscle mass. Unfortunately, taking testosterone can also cause depressed sperm production, elevated bad

PLENTY OF GOOD COMPANY
At Sun City, retirement is seen as a long whirl of activity, and residents exude an extraordinary sense of well-being. Above right, Arizona's Sun City Poms, whose members range in age from 62 to 78; right, the Desert Aires Barbershop Chorus. Founded in 1960, the Arizona town was the first Sun City; there are now more than 50.

MAN OF STEEL *Jeffry Life has become a poster boy for Cenegenics, whose program features exercise, diet and—controversially—hormone supplements.*

cholesterol, shrunken testicles, water retention and bad skin. (I declined Leake's prescription.)

As to growth-hormone supplementation, the jury is still out. A 1990 study of its effects on a group of men ages 61 and above, published in the *New England Journal of Medicine*, was broadly favorable. Subjects showed marked increases in lean body mass and no apparent side effects. But it was a small study, of short duration. In 2003, the *Journal* returned to the subject. This time the conclusions weren't as encouraging. Growth hormone changes body composition but doesn't appear to improve function. And there are niggling concerns that it could accelerate the growth of cancer, the article noted. The alternative? "Going to the gym is beneficial and certainly cheaper."

Biology Is Destiny

There is only one documented way to lengthen life—caloric restriction—and that hasn't been conclusively proven to work for humans. No wonder there was so much excitement around a compound called resveratrol, which seems to mimic the effects of such a diet without the need for a punitive regimen. David Sinclair, a Harvard Medical School professor of genetics, has spearheaded research on resveratrol, which is found in red

grapes and activates sirtuins, enzymes involved in regulating metabolism. In 2008 Sinclair sold his company, Sirtris Pharmaceuticals, to GlaxoSmithKline for $720 million. But that investment has yet to pay off. A study of one Sirtris resveratrol drug was stopped because of safety concerns, and the scientific community is divided over the potential of the research.

One problem with possible elixirs of youth is the danger that they may indeed stop aging—by killing the patient. Research into an enzyme called telomerase, which appears to "immortalize" cells by lengthening the telomeres—the repetitive sequences of DNA at the end of each chromosome, which shorten when cells divide—has shown promise. But high-strength compounds "are believed to be too toxic for human use. We need to do a lot of medicinal chemistry before we can get them to preclinical trials," e-mails Jon Cornell of Sierra Sciences, a Nevada-based biotech company focused on telomere research. In the meantime, both resveratrol and telomerase are available in weaker formulation as dietary supplements.

As Young as They Feel

"Strictly speaking, longevity is measured in numbers: it is the arithmetical accumulation of days, weeks, months and years that produces our chronological life," wrote the late psychiatrist and gerontologist Robert Butler in his book *The Longevity Prescription*. "Yet aging—or, more accurately, its converse, staying young—is in no small measure a state of mind that defies measurement."

That isn't a platitude, as Harvard psychology professor Ellen Langer set out to prove in 1979. Her experiment started with the retrofit of an isolated hotel in New England. The fixtures were exchanged for 1959 period equivalents; the refrigerator was stocked with foodstuffs available 20 years earlier. Then came the guests: men in their 70s and 80s, instructed to pretend they had traveled back two decades in time.

This pretense proved decisive. A control group, taken from the same demographic, arrived to stay in the hotel after the first contingent had left. Their experience differed in only one respect: they were allowed to acknowledge that this was an experiment. In just a week, both groups chalked up physical and cognitive improvements. But the changes were much more pronounced among the time travelers.

It's a result that goes some way toward explaining the sense of well-being that radiates from the residents of Sun City Shadow Hills, a community for "active adults aged 55 and better" in Southern California's Coachella Valley, where retirement has been reimagined not as a cessation of work but as a long whirl of absorbing activity.

A construction magnate named Del Webb opened the first Sun City community in Arizona in 1960. "An old fellow came up to me once with tears in his eyes and thanked me for building Sun City," recounted Webb, interviewed for a 1962 TIME cover story. "He said he was planning to spend the happiest 40 years of his life there." Webb died in 1974, but his creation lives on. There are now more than 50 Sun Cities dotted across America.

Sun City Shadow Hills, launched in 2004, is among the newest. Driving along its flawless streets beneath unrealistically blue skies, you wonder if you've strayed onto a backlot at a Hollywood studio. Distant figures shimmer in the heat haze on the fairway, but the sidewalks are empty. Then you ring the doorbell at Patti and Phil Wolff's house and discover a fair chunk of Sun City's population in their large kitchen. A barbecue sends smoke signals to the rest of the community that one of Patti's famous meals will shortly be served. Phil, 63, plays softball, is a regular at the state-of-the-art gym at Sun City's Montecito Clubhouse, and indulges a passion for cycling, racking up as many as 175 miles per week. "Basically, he does so many things that I hardly see him," says Patti, 62. "It's like if he was back at work again."

Several of the friends around their table still work.

By 2050, more than a fifth of humanity will be

60

or older; in the U.S. the 60-pluses will make up

27%

of the population.

Larry Johnson, 63, once the manager of a funeral home in Oregon, these days offers a similar service for pets, assisted by his partner, Bruce Atkinson, 66. "Hundreds of people [in the area] have old dogs and cats, and eventually those dogs and cats will die, so Larry gets a lot of business," says Atkinson.

The same forces that keep Johnson's business ticking will eventually disrupt the idyll that he and his friends have built. On a recent night out at another Sun City development called Palm Desert, Johnson and Atkinson caught an uncomfortable glimpse of the future. "We were standing there, and we both made the comment 'God, these are really old people,'" says Atkinson. Palm Desert opened 12 years before Shadow Hills; the average age of its residents is higher. And because ownership of Sun City properties is restricted to the 55-pluses, the communities age and risk dying off together. Yet the absence of young people helps—for a while, at least—to revitalize Sun City residents, permitting an illusion of agelessness.

Keeping On Keeping On

The fastest-growing segment of the world population is the very old, with the number of centenarians projected to reach nearly six million by 2050. But as John F. Kennedy observed in a 1963 address to Congress, "It is not enough for a great nation merely to have added new years to life. Our objective must be to add new life to those years." Life spans have lengthened; health spans have not kept pace. Genes and luck play a role in how we age. Lifestyle and the wealth required to enhance it are also key factors. Science may yet devise an elixir that allows us all to be Mick Jaggers, doing what we like, seemingly indefinitely. But the swelling ranks of amortals may find themselves instead subsumed into another phenomenon of our times: the living death before death, sometimes lasting decades, that increased longevity without extended vitality represents.

Still, the trend of amortality is accelerating. You can't just close your eyes and wish us back in Kansas among kindly folk who obligingly conform to outdated expectations of age. Look around our virtual Vegases, and you'll see cause for optimism too. Amortals hold the key to transforming perceptions by showing what older people can do and showing older people what they can be. They're inclined to keep working, rather than vegetating. They may not age gracefully, but neither do they trade their sense of adventure for dignity. Thanks to amortality, our graying world may not prove too gray a place.

Adapted from Catherine Mayer's book Amortality: The Pleasures and Perils of Living Agelessly.

Looking Good

THE "FIT VS. FAT" DEBATE
EXERCISE ALONE WON'T MAKE YOU THIN
WHAT'S AGING YOUR SKIN
WHY WE'RE SPENDING SO MUCH ON BOTOX
12 THINGS THAT CAN RUIN YOUR SMILE
HAIR TODAY, AND TOMORROW

WHILE FEELING YOUNG IN MIND AND SPIRIT IS great, most of us would still like to look our age—well, maybe slice a few years off that number. Thankfully, science is on our side here too. Research is making clearer what really helps keep extra pounds at bay (diet is much more important than tons of time at the gym, it turns out) and how to keep your metabolism revved up so you're burning as many calories as possible (studies say that how early in the day you eat, how you exercise, and the amount of good-quality sleep you get are all key). And since nothing gives away our age like skin—the body's biggest organ—we're getting a better idea of what truly minimizes wrinkles (skip the sweets and get stress under control); what inject-ables, chemical peels, and cosmetic fillers can do; and how to keep more of our hair—or learn to let it go.

The "Fit vs. Fat" Debate

Extra pounds can carry plenty of health risks. But weight, it turns out, may not be the only harbinger of poor health.

✳ BY ALICE PARK

F YOU'RE LIKE MOST PEOPLE, YOU'RE ALL too familiar with how much you weigh. But how many of us know how *fit* we are? And does fitness really matter when it comes to our health?

With about a third of the U.S. population considered obese, and over 69% who are either overweight or obese, it's not surprising that a small but vocal group of folks who tip the scales beyond their recommended body weights are making the case for being heavy but healthy. Their argument: if their blood pressure and heart function are in normal ranges, why should their weight matter?

It's certainly true that carrying extra pounds can significantly increase the risk of developing chronic conditions including diabetes, high blood pressure and heart problems. That's why doctors and public-health officials urge people to maintain a healthy weight—to lower those risks. But there may also be something to the position that weight, or gaining it, may not be the only sign of ill health.

Take, for example, a recent study that tried to tease apart the role that fitness plays in longevity. Most studies that have previously linked weight gain, overweight and obesity to a higher risk of dying early have focused only on BMI, or body-mass index, a ratio of height and weight. That's because weight can indirectly affect a number of different metabolic processes that contribute to our longevity, such as how we burn calories or process sugars and how high our blood pressure is. But weight may also be masking the effect of another factor that could either protect us or propel us to an early death: how efficiently our hearts and lungs are working—or, in other words, how fit we are.

Duck-chul Lee, assistant professor at Iowa State University, conducted a study as a postdoctoral fellow at the University of South Carolina's Arnold School of Public Health to focus specifically on the role of fitness in overall mortality rates as well as deaths due to heart disease. He and his colleagues recruited a group of more than 14,000 middle-aged men, most of them white and upper middle class, for a long-term study. Lee had the men run on a treadmill to measure their heart and lung functions at two points during the trial. He compared their maximum fitness levels—how long they could run at increasingly steeper inclines—with death rates

across the group. He also factored in the men's weight.

The guys who maintained their fitness levels between the times the two measurements were taken lowered their risk of dying from heart-related or any other causes by up to 30%, compared with men who had lost fitness. Those who actually improved their fitness lowered their risk of dying even more, by up to 44%. In fact, for every unit improvement in fitness—this was measured as metabolic equivalents (METs), or roughly how much energy it takes to complete specific physical activities—there was a 15% decrease in death from any cause, and a 19% decrease in dying from heart-related events.

The surprising part was not that the men who were in better shape lived longer than those who weren't. The real shocker came when Lee and his colleagues realized that all of these changes occurred *regardless of how much weight the men gained or lost*. When it came to BMI, fluctuations during the 11 years of the study weren't linked to any changes in all-cause mortality, though men whose BMI went up had an increased risk of dying from a

heart event compared with those whose BMI went down. "Regardless of weight change—some lost weight and some gained, while some remained stable—loss of fitness was associated with a higher risk of mortality," says Lee.

It's a confusing concept. After all, isn't weight a reflection of how fit we are? Well, yes and no. To a certain extent, it's true that the more weight we gain, the less fit we tend to be. In fact, when Lee and his team looked at which participants in the study lost fitness, those were the sedentary men who started smoking and developed conditions such as diabetes and hypertension. And the men who were the least physically active also lost the most fitness.

It's easy to assume that a high BMI is primarily due to extra fat tissue, and in many cases it is. But muscle also contributes to a person's heft, and people who are more active are likelier to develop more muscle tone, which may add to their weight—and their BMI—without necessarily harming their health. That's why the researchers wanted to distinguish fitness from weight to pinpoint how each contributes to someone's risk of dying. "When you change your body weight, you have to consider whether you become more fit or not," says Lee. "If you gain weight but become more fit, then that might be OK regarding your mortality risk. We have to start considering other factors when

69

we talk about weight change and health outcomes."

He may have a point. In another analysis involving 18,670 healthy men and women in their 40s and 50s, those who were more fit in middle age were less likely to experience conditions such as diabetes, stroke, lung cancer and even Alzheimer's disease a couple of decades later. These participants also enjoyed fewer chronic conditions in the last five years of life, meaning they spent more of this time healthy rather than burdened by disease. "The results show that fitness can not only delay [disease] but prevent it," says the study's lead author, Dr. Jarett Berry of the University of Texas Southwestern Medical Center. Healthier years—not just more years—is what we're all hoping for, right?

A group of scientists at the University of Vermont found a similar benefit in warding off cancer among a group of 17,000 men enrolled in the Cooper Center Longitudinal Study in Dallas. Each of the men agreed to take a treadmill test that measured their maximum oxygen intake, a stand-in measurement for how fit they were. The fitter men were less likely to develop lung or colon cancers after 20 years, compared with those who weren't in good shape. And more important, this trend didn't seem to be affected by how much they weighed.

> "If you gain weight but become more fit, that might be OK," says one researcher. "We have to consider other factors when we talk about weight and health."

Men who were normal weight but not fit had a higher risk of getting cancer than those who were fitter.

But what if you're already in your 40s and aren't in the same physical prime that you maintained in your youth? There's evidence that it's not too late to whip yourself into shape to reap these health benefits, according to another study from the University of Texas. In that trial, which involved 9,000 middle-aged men and women, those who improved their fitness levels after 18 years had fewer Medicare claims for heart-failure treatments than those who didn't improve over that period. Reassuring news, indeed.

All of this doesn't necessarily mean that weight does not play a role in health, especially for those who are obese. In Lee's study, for example, most of the participants clocked in at close to normal weight or only slightly overweight. Other studies show that among the obese, weight loss can have a much more dramatic effect in lowering the risk of dying from heart events or other causes.

So reducing your risk of an early death may be more complicated than simply watching the scale. If you're trying to stay healthy and to protect yourself from a chronic condition, you might not need to shed pounds, but you'll still have to exercise. And that hasn't changed—the best way to stay fit is to be physically active.

Meet the World's Oldest Female Bodybuilder

BY ALEXANDRA SIFFERLIN

Now 77 years old, Ernestine Shepherd wasn't always so ripped and fast. She got interested in bodybuilding when she was 56, with her sister, Mildred. When Mildred died shortly afterward from a brain aneurysm, Shepherd held onto their dream, and in 2010 she was crowned the World's Oldest Performing Female Bodybuilder by Guinness World Records. We spoke to Shepherd, who lives in a suburb of Washington, D.C., about her daily routine and what drives her to stay in shape.

What keeps you motivated?
I made a promise to [my sister] that I would follow her dream, and it has become mine. She said that we wanted to inspire and motivate others to live a healthy, happy and fit lifestyle, to let them know that age is nothing but a number and you can get fit. After my sister died, I ended up with high blood pressure, panic attacks, high cholesterol—you name it, I had it. After a lot of prayer and help from my family, I could get on my feet again, and I started running. I found I didn't need to take all that medication I was taking. My blood pressure went down, I stopped feeling unhappy, I stopped feeling depressed.

Is there anyone you look up to for motivation?
You will laugh when I tell you. I just love some Sylvester Stallone. I was inspired by the Rocky Balboa character, and I found it was not about how hard you can get hit but how you keep moving forward. That is how winning is done. The *Rocky* movies make me ready to take on the world. My mantra is: "Determined, dedicated, disciplined to be fit."

How did you get into bodybuilding specifically?
When my sister and I began working out, she said, "We are going to be two of the oldest competitive bodybuilders." She said we were going to make the *Guinness Book of World Records* by the end as two sisters. Before she died, she looked at me and said, "If I don't make it, you have to continue what we started." I met the former Mr. Universe Yohnnie

God, they were a mess. We were complete opposites.

Do you have any guilty pleasures?
I really don't, because I have acid reflux and I have to be mindful. But I don't take any medication for that, because I know what foods trigger it for me. You know what I like? I like wearing skimpy clothes. Oh, yes, I do. I love to have my back out and my stomach out. I like the catsuits—all of those things. I couldn't wear them years ago.

What advice do you have for people 50-plus to get in shape?
I would tell people ages 50 and older who haven't done any type of exercise before to start out very, very slowly. Don't jump in with both feet because you don't want to stop. Take your time and find out what you like to do. Find what you like and stick with it. But remember, no matter what you do, you have to be determined. You have to be dedicated and disciplined. It takes four things to get fit: eat correctly, drink plenty of water, do some type of strength training, and do some kind of cardio. Don't listen to anyone who says you're too old or starting out too late. Once you start exercising, honey, it's a whole new ballgame.

A Day in the Life of an Extraordinary Septuagenarian

2:30 a.m. Wake up. Meditate and read devotions from the Bible. Eat a snack of a bagel with peanut butter and hard-boiled egg whites. Drink 16 ounces of water.

3:45 a.m. Head to nearby park and run 10 miles. Eat breakfast of oatmeal, three hard-boiled egg whites, and a tablespoon of walnuts. Drink eight ounces of liquid egg whites.

8 a.m. Head to the gym and work out for an hour and 45 minutes.

10 a.m. Train a group of senior men and women. The oldest woman is 89 years old.

11 a.m. Train four to five women in the gym. Drink another eight-ounce glass of liquid egg whites.

1 p.m. Go home and eat a can of tuna, a cup of spinach and half a cup of sweet potato; drink an eight-ounce glass of water. Rest.

6 p.m. Teach another class at the gym. Head home and eat turkey, brown rice, broccoli and more egg whites; drink lots of water.

10–10:30 p.m. Drink one more glass of liquid egg whites. Go to bed.

Shambourger, and I asked him if he would work with me because I wanted to become a bodybuilder. I was 71 years of age when we started. He said, "You are going on a long journey, and you are going to have to follow everything I tell you to do. Do you think you can do it?" I shook my head and said yes. In a matter of seven months he had me ready to be on the stage. We haven't looked back since then.

You were 56 when you and your sister first started working out. What was your perspective on fitness at that time?
I was such a prissy woman. I didn't want to do any exercising. My sister was very active. When she started working out, she was 99 pounds and skin and bones. She had to gain enough weight to meet her goal of 140 pounds, which she did. I was 145 pounds. I had to come down in weight because I had the cellulite, I had the fat in the back, the legs were—oh, my

Exercise Alone Won't Make You Thin

You've heard it for years: to lose weight, hit the gym.
But while physical activity is crucial for good health,
it doesn't always melt pounds—in fact, it can add them.
Here's why.

✳ BY JOHN CLOUD

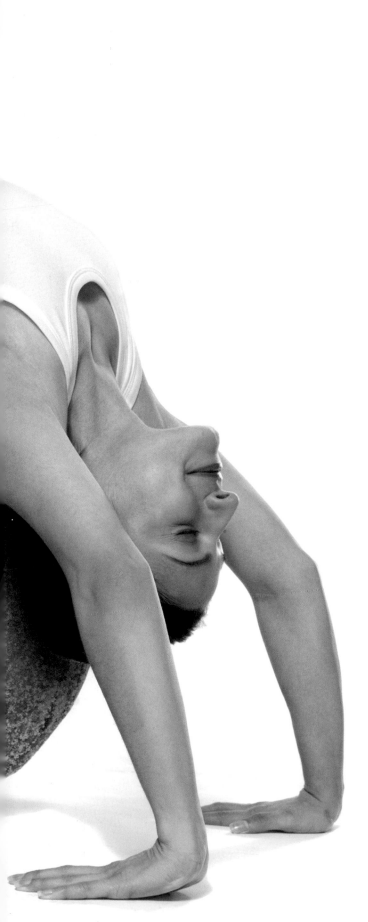

A S I WRITE THIS, TOMORROW is Saturday, a major weight-lifting day. I'll spend five minutes warming up on a treadmill until my heart beats 150 times per minute. Then I'll do three sets each of back, shoulder, biceps, triceps and leg exercises, some of them combined (so that, for example, I hold 15-pound barbells at my ears, lunge down, and then rise and push the barbells as high as I can). On other days I do an hour of cardio (40 minutes on the treadmill, 20 minutes on a bike), and then on *other* days I work my core muscles with mat exercises like the plank, holding myself in a stable push-up position for at least 95 seconds, those final five seconds a grueling expiation of any gastronomic indulgences during the week.

I have exercised this way—obsessively, a bit grimly—for nearly 20 years. But a few years ago, when I first wrote about the topic, I began to wonder why. Even after all those hours at the gym, I can't seem to lose the fat that hangs over my belt when I sit. In fact, as I age (I'm now 42), I gain a little each year. Given the number of sit-ups I can do, my stomach should look like a steel plate. Instead I have to suck it in every time I try on new jeans. Why isn't the exercise working?

It's a question a lot of us could ask. Some 50 million Americans now belong to a health club, up from 23 million in 1993. We spend something like $25 billion a year on gym memberships. Yet obesity figures have risen dramatically in the same period: approximately 35% of Americans are obese, and another third count as overly fat (as I do) when you use basic measures such as the body-mass index, which accounts for height, weight and age. Yes, it's entirely possible that those of us who regularly work out would weigh even more if we exercised less. But like many other people, I get hungry after I exercise, and I often eat more on the days I work out than on the days I don't. Could exercise actually be *keeping* me from losing weight?

The conventional wisdom that exercise is essential for shedding pounds is actually fairly new. As recently as the 1960s, doctors routinely advised against rigorous physical activity, particularly for older adults who could injure themselves. Today doctors encourage even their oldest patients to exercise, which is sound advice for many reasons: people who do so regularly are at significantly lower risk for all manner of diseases, those of the heart in particular. They less often develop cancer, diabetes and other health problems, including neurological decline. According to the latest research, diseases such as dementia can be slowed when patients repeatedly do exercises such as balancing on one leg while lifting a ball in the opposite hand. That transverse motion may stimulate neural pathways that are otherwise underused.

But the past few years of obesity research show that the role of exercise in weight loss has been wildly misunderstood. "In general, for weight loss, exercise is pretty useless," said Eric Ravussin, director of the Nutrition Obesity Research Center at Pennington Biomedical Research Center in Louisiana. The basic problem is that although it's true that physical activity burns calories—and that you must burn calories to lose weight—exercise has another effect: it can stimulate hunger. That causes us to eat more, which in turn can negate the weight-loss benefits we just accrued. Exercise, in other words, isn't necessarily helping us lose weight. It may even be making it harder.

The Compensation Problem

In 2009, the peer-reviewed journal *PLOS ONE* published a remarkable study of 464 overweight, sedentary women broken into four groups. Those in three of the groups were asked to work out with a personal trainer for 72, 136 and 194 minutes per week, respectively, for six months. Women in the fourth cluster, the control group, were told to maintain their usual physical-activity routines. All were asked not to change their dietary habits and to fill out monthly medical-symptom questionnaires.

The findings were surprising. On average, the women in all four groups lost weight, but those who exercised—sweating it out with a trainer several days a week for six months—did not lose significantly more than the control subjects did. (The control-group women may have lost weight because they were filling out those health forms, which may have prompted them to think more carefully about what to eat.) Some in each group actually gained weight, some more than 10 pounds each.

What's going on here? Dr. Timothy Church, professor of preventive medicine at Pennington, called it "compensation." You and I might know it as the lip-licking anticipation of perfectly salted, golden-brown french fries after a hard trip to the gym. Whether because exercise made them hungry or because they wanted to reward themselves (or both), most of the women who exercised ate more than they did before they started the experiment. Or they compensated in another way, by moving around a lot less than usual after they got home.

In 2011, the American College of Sports Medicine issued guidelines stating that at least 250 minutes of exercise per week provides "clinically significant" weight loss. As an ACSM spokesman pointed out, that's just 35 minutes a day. It doesn't sound bad, but there are hidden costs—you drive to the gym, park, check in, change, fumble with your padlock, work out for those 35 minutes, shower, step around someone else's wet towels, then drive home. Well more than an hour has vanished from your day. And don't you deserve a latte on the way home?

The problem with strict exercise regimens is that, for many people, they lead to ravenous compensatory eating. Church's data show this clearly: we indulge ourselves

Burning Calories

What it takes for a 154-pound, 30-year-old woman to work off a blueberry muffin.

Lawn-mowing
66
MIN.

Skating (fast)
21
MIN.

Gardening
66
MIN.

Lifting weights
115
MIN.

Source: U.S. Department of Agriculture Center for Nutrition Policy and Promotion

after we work out. So if you run for 35 minutes a day but regularly stop for a latte afterward, you may not realize that your jog burned only 200 to 300 calories and that a 16-ounce Starbucks caffè latte with skim milk has 130 calories. If you add a pastry as a treat, it would have been better not to have left your house in the first place.

And yet: doesn't exercise turn fat to muscle, and doesn't muscle process excess calories more efficiently than fat? Yes, although the muscle-fat relationship is often misunderstood. According to widely used calculations published in the journal *Obesity Research* by a Columbia University team in 2001, a pound of muscle burns approximately six calories a day in a resting body, compared with the two calories that a pound of fat burns. Which means that after you work out hard enough to convert, say, 10 pounds of fat to muscle—a major achievement—you would be able to eat only an extra 40 calories

Cycling (easy pace)
77
MIN.

Vacuuming
92
MIN.

Jogging (5 mph)
33
MIN.

Folding laundry
230
MIN.

Blueberry muffin
360 calories

per day, about the amount in a teaspoon of butter, before beginning to gain weight.

Another problem is that fundamentally, humans are not a species that evolved to dispose of calories beyond what we need to live; we almost instantly store most of the calories we don't need in our fat cells. All this helps explain why our herculean exercising over the past 30 years—the personal trainers, the gleaming new gyms with high price tags, the "fitness centers" added to thousands of hotels—hasn't made us thinner.

Self-Control Is Like a Muscle
Many people assume that weight is mostly a matter of willpower, that we can learn both to exercise and to avoid lattes. A few of us can, but evolution did not build us to do this for long. In 2000, the journal *Psychological Bulletin* published a paper by psychologists Mark Muraven and

Roy Baumeister in which they observed that self-control is like a muscle: it weakens each day after you use it. If you force yourself to jog for an hour, your self-regulatory capacity is proportionately enfeebled. Rather than lunching on a salad, you'll be more likely to opt for pizza.

Some of us can will ourselves to overcome our basic psychology, but most of us won't be very successful. "The most powerful determinant of your dietary intake is your energy expenditure," said Steven Gortmaker, who heads Harvard's Prevention Research Center on nutrition and physical activity. "If you're more physically active, you're going to get hungry and eat more."

In 2008, the *International Journal of Obesity* published a paper by Gortmaker and Kendrin Sonneville of Boston Children's Hospital noting that "there is a widespread assumption that increasing activity will result in a net reduction in any energy gap"—*energy gap* being the term scientists use for the difference between the number of calories you use and the number you consume. But Gortmaker and Sonneville found in their 18-month study of 538 students that when kids start to exercise, they end up eating more—not just a little more, but an average of 100 calories more than they had just burned.

Part of the problem could be simply that we're pushing people to exercise too much, too vigorously. At my Equinox gym in New York City—a gleaming club with three floors, glass-encased saunas and a fingerprint system for admittance—the "30/60/90" class involves such difficult repetitions (for 30, 60 and then 90 seconds) that I have seen people collapse in the back of the room; others scream but continue. Fortunately, there's a juice bar in the front that can satisfy any craving for post-workout sugar. The gym makes out both ways.

In short, all this "fitness" could be adding to our obesity problem. Because exercise depletes not just the body's muscles but also the brain's self-control "muscle," many of us feel entitled to eat a bag of chips during that lazy time after we get back from the gym. This explains why exercise could make you heavier—or at least why even my wretched hours of effort each week aren't eliminating all my fat. It's likely that I'm more sedentary during my nonexercise hours than I would be if I didn't work out with such puritan fury. If I exercised less, I might feel like walking more instead of hopping into a cab; I might have enough energy to shop for food, cook and then clean instead of ordering a satisfyingly greasy burrito.

Closing the Energy Gap
The problem ultimately is about not exercise itself but the way we define it. Many obesity researchers now believe that very frequent, low-level physical activity—the kind humans did for tens of thousands of years before the leaf blower was invented—may actually work better than the occasional bouts you get at a place like Equinox. "You cannot sit still all day long and then have 30 minutes of exercise without producing stress on the

muscles," said Hans-Rudolf Berthoud, a neurobiologist at Pennington who has studied nutrition for more than 30 years. "The muscles will ache, and you may not want to move after. But to burn calories, the muscle movements don't have to be extreme. It would be better to distribute the movements throughout the day."

I was skeptical when Berthoud said this. Don't you need to raise your heart rate and sweat in order to strengthen your cardiovascular system? Don't you need to push your muscles to the max in order to build them? Actually, it's not clear that vigorous exercise like running carries more benefits than a moderately strenuous activity like walking while carrying groceries.

There's also growing evidence that when it comes to preventing disease, losing weight may be more important than improving cardiovascular ability. In 2009, Northwestern University researchers released the results of one of the longest observational studies ever to investigate the relationship between aerobic fitness and the development of diabetes. The results? Being aerobically fit was far less important in preventing the disease than having a normal body-mass index.

So how did the exercise-to-lose-weight mantra become so ingrained? Public-health officials have been reluc-

Many obesity researchers now believe that very frequent, low-level physical activity may actually work better for us than occasional bouts of exercise at a gym.

tant to downplay exercise because those who are more physically active are, overall, healthier. Plus, it's hard even for experts to renounce the notion. For years, psychologist Kelly Brownell ran a lab at Yale that treated obese patients with the standard, drilled-into-your-head combination of more exercise and less food. Only about 5% of participants could keep the weight off, and although those 5% were more likely to exercise than those who got fat again, Brownell told me that if he were running the program today, "I would probably reorient toward food and away from exercise."

Some research has found that the obese already "exercise" more than most of the rest of us. In 2009, Dr. Arn Eliasson of Walter Reed Army Medical Center reported the results of a small study that found that overweight people actually expend significantly more calories every day than people of normal weight—3,064 versus 2,080.

In short, it's what you eat, not how hard you try to work it off, that matters more in losing weight. You should exercise to improve your health, but be warned: fiery spurts of vigorous exercise could lead to weight gain. I love how exercise makes me feel, but tomorrow I might skip all those lunges—and the berry smoothie that is my post-exercise reward.

A Faster Metabolism at Any Age

BY JULIA SAVACOOL

SPEEDING UP YOUR METABOLISM IS OUT of your control, right? Not quite. Although genetics and age play roles, recent studies suggest that you have plenty of say over how well your metabolism functions. In fact, you can all but negate the metabolic slowdown that happens after 40 by tweaking your diet, exercise and sleep habits. "Think of your body as an engine—your metabolism is the rate at which your engine runs," explains Dr. Scott Isaacs, author of *Hormonal Balance: How to Lose Weight by Understanding Your Hormones and Metabolism.* "By making adjustments to these three elements, you can actually make your engine rev higher."

1. Eat Early Your basal metabolic rate—the number of calories your body burns at rest—is based on things like age, height and body type, so there isn't much you can do to alter it. But there is a lot you can do to change the number of calories you burn above that. Specifically, eat breakfast. Leaving for work on an empty stomach is like hitting the pause button on your metabolism. Here's why: when your brain senses that your stomach is empty, it sends a message to your cells to conserve energy in case another meal doesn't arrive. In other words, your body holds onto the fat stored in your cells instead of helping you burn it off.

2. Eat Often Dr. Mark Hyman, author of *The Blood Sugar Solution,* recommends eating small meals every three or four hours. Aim to make each meal at least one-quarter protein, whether animal protein, beans or dairy, says Marissa Lippert, a registered dietitian. A 2011 study in the journal *Neuron* suggests that protein stimulates the cells responsible for switching on the body's calorie-burning mechanism. Foods high in sugar and processed carbs, on the other hand, can lead to insulin resistance, causing your body to store extra calories as fat.

3. Sweat Off the Weight "Not only does exercise affect your metabolism while you're doing it, but research shows you can keep burning calories up to 24 hours after you finish because your metabolism stays elevated," says Isaacs. That's especially true if you challenge yourself: a study in the journal *Cell Metabolism* shows that intense bouts of exercise can "turn on" genes responsible for energy metabolism. Activation of these fat-burning genes was higher in cyclists who pedaled at 80% of aerobic capacity than in those who did more-moderate cycling. (Caveat: for most people, vigorous exercise stimulates appetite, so keep an eye on calories. Working out isn't a license to eat.) Exercise is particularly helpful once you pass the age of 40, when metabolism naturally begins to slow down. A study in the journal *The Physician and Sportsmedicine* found that fit women ages 41 to 81 who continued to exercise four to five times a week as they got older had little change in body composition. The real reason you lose muscle with age? You stop using it.

4. Sleep Away the Pounds Research indicates that people who sleep two-thirds of their usual amount (five hours instead of eight, say) eat an average of 549 extra calories the following day. Experts believe that getting too few Z's upsets the balance of appetite-regulating hormones. "Resistance to leptin—a hormone that regulates body weight—increases, while levels of ghrelin, a hormone that signals to your brain that you're hungry, also increase," explains Isaacs. Aim for seven to eight hours of pillow time, advises Hyman: "Just a small change in your sleep schedule can make a big difference in your health." Not to mention your ability to burn calories.

This article was originally published in Health *magazine.*

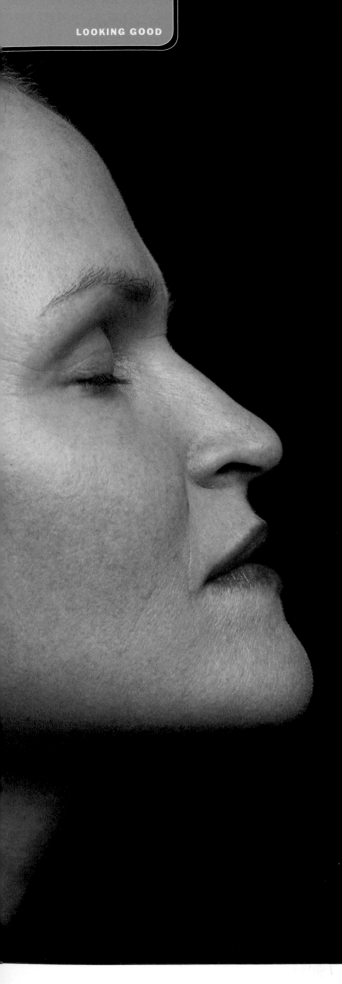

What's Aging Your Skin

It goes way beyond genes and the sun.
Avoiding these eight common pitfalls
will help keep your skin youthful.

 BY STACEY COLINO

SKIN AGING, LIKE MOST PHYSI-ological phenomena, is the result of many things. "But only about 20% to 30% of the process is genetically determined," says Dr. Doris J. Day, a clinical associate professor of dermatology at New York University Medical Center. "So there's more you can do to control it than you might have thought." We know you already wear sunscreen every day to fend off those UV rays; here are some stealthier skin agers and ways to mitigate their effects.

Sweet Treats If sugary foods are a staple of your diet, you may want to reconsider what you eat. "When sugar breaks down and enters the bloodstream, it bonds with protein molecules, including those found in collagen and elastin [the fibers that support skin], through a process called glycation," says Dr. Leslie Baumann, a dermatologist in Miami Beach. "This degrades the collagen and elastin, which in turn leads to sagging and wrinkles."
PREVENTIVE MEASURES: Curb your consumption of simple carbohydrates, which include the obvious treats, like soft drinks and candy, but also seemingly innocuous choices such as honey, white rice and white bread. These foods are quickly converted into sugar in your body and put your skin on the fast track to glycation. If you need something sweet (and really, who doesn't?), Baumann suggests a small square of dark chocolate. The antioxidants in it can protect you from free radicals, those unstable atoms in the atmosphere that latch onto skin and lead to fine lines. Also, increase your intake of vitamin

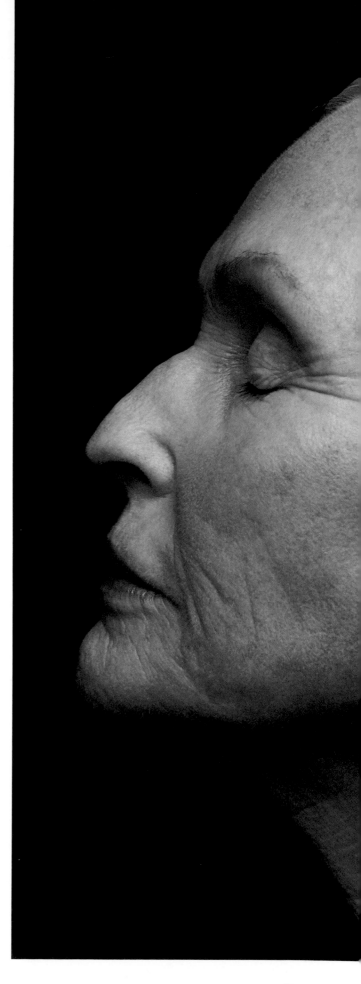

C. "It helps generate collagen," says Dr. Ellen Marmur, a dermatologist in New York City and the author of *Simple Skin Beauty.* You'll find vitamin C in papayas, strawberries, broccoli, oranges and kiwis.

Frequent Flying You're much closer to the sun in a plane than on land, so it stands to reason that solar rays, which can penetrate windows, "are more intense at higher altitudes," says Marmur. This may explain why pilots and flight attendants have been found to be at an increased risk for melanoma and other skin cancers. Plus, the air up there is notoriously dry—and without moisture, skin, like any other living tissue, simply shrivels.

PREVENTIVE MEASURES: Drink as much water as you can in flight; avoid alcohol, which is dehydrating; and apply a rich moisturizer with SPF 15 or higher 30 minutes before boarding, as sunscreen needs time to be absorbed before it's effective. And if you're sitting next to a window, pull down the shade.

Midlife Moisture Loss For women who are in menopause, "your body begins to pump out less estrogen," says Dr. Arielle Kauvar, a dermatologist and the director of New York Laser & Skin Care in New York City. "Since estrogen stimulates oil and collagen production in the skin, your skin may become drier, more wrinkled, and saggy as estrogen levels drop."

PREVENTIVE MEASURES: Your best inexpensive bet may be to "troubleshoot by moisturizing heavily," says Marmur. For women who are taking hormone-replacement therapy, these drugs can offset some of the effects of moisture loss, but it can take time to find the exact combination of hormones that your body will respond

DO: Get plenty of sleep

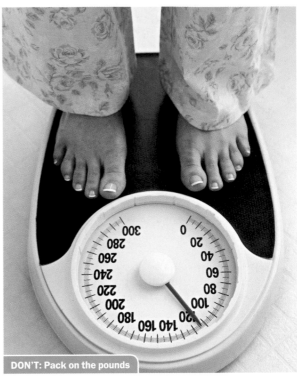

DON'T: Pack on the pounds

FROM LEFT: DANA HOFF/BEATEWORKS,/CORBIS; MICHAEL A. KELLER,/CORBIS; FOODCOLLECTION/GETTY IMAGES; ANTOINE ARRAOU/PHOTOALTO/CORBIS

to. For instant results, you might consider dermal fillers, which are injected into the skin to smooth wrinkles and plump up sagging areas. Consult your dermatologist for options.

Weight Fluctuations Packing on pounds can make your skin look more filled out on the surface, but carrying excess weight can cause a rise in your body's levels of the hormones insulin and cortisol, which can weaken supportive collagen fibers. "You'll see increased sagging from putting and keeping on as little as 10 to 15 extra pounds," says Dr. Fredric Brandt, a dermatologist in Miami and New York. In addition, repeatedly gaining and losing weight can take its toll on the skin's elasticity, leaving behind stretch marks and jowls.
PREVENTIVE MEASURES: Aim to keep your weight in the normal range, with a body-mass index of between 18.5 and 24.9.

Untamed Tension Yes, stress really does wear on you. When you're under intense or chronic pressure, your body increases production of cortisol, which can decrease the skin's ability to repair itself. What's more, stress can make you tense up and grimace or frown, often without your even realizing it. After a while, these repeated muscle contractions can leave their mark in the form of permanent lines, says Brandt.
PREVENTIVE MEASURES: Get stress and anxiety under control by exercising regularly. Yoga, tai chi and brisk walking have been found to be effective tension tamers,

possibly because of the meditative aspect of these activities. (The deep or rhythmic breathing of yoga and tai chi probably helps, too, by promoting healthy circulation.) If you don't have time for hourlong exercise sessions, break up your workouts: walk the 30 minutes to the office, and download a yoga app (such as Yoga RELAX) so you can do a few gentle poses before bed.

The Blues Depression doesn't show up only in your demeanor—it may also show up on your face. Not only can a frown (ironically, just like a smile or a squint) become permanently etched into the skin, but depression is also associated with "a decrease in growth-hormone synthesis, which inhibits the ability of the skin to repair itself at night," says Brandt. And when people are depressed, they may not take care of themselves (or their complexions) the way they should.
PREVENTIVE MEASURES: To combat depression, exercise regularly, enter counseling if necessary, and speak to your doctor about whether you would benefit from an antidepressant. Interestingly, reducing wrinkles with a cosmetic treatment like Botox might help ease symptoms of depression. Sure, that could just be because looking better can make you feel better, but a study conducted at Cardiff University in the U.K. found a more likely explanation: when people had their frown lines treated with Botox, the paralysis of those facial muscles prevented them from transmitting negative-mood signals to the brain, which correlated with a lifting of the spirits.

DON'T: Eat the sweet stuff

DO: Exercise regularly

Lack of Sleep "Your skin has a chance to repair itself overnight," says Dr. Mary Lupo, a clinical professor of dermatology at Tulane University in New Orleans. "Without enough deep sleep, the kind that you can't be roused from easily, the skin can't properly undo daily damage." Also, sleep deprivation puts your body into stress mode, which causes it to release more stress hormones.

PREVENTIVE MEASURES: Shoot for seven to eight hours of sleep a night. It takes discipline, but start by shutting off all electronic devices a half-hour before bed so the stimulation doesn't keep you up. Another strategy: try to sleep on your back. "If you usually sleep with your face smooshed into your pillow," says Lupo, "it will look creased faster."

Marathon Workouts In the skin-aging equation, regular moderate exercise is a plus, since it reduces stress. But if you frequently run or bicycle long distances outside, you are not only exposing yourself to lots of UV light but also "jolting, and possibly damaging, the support structure of the skin," says Brandt.

PREVENTIVE MEASURES: Do not consider this a license to slack off! Brandt emphasizes that premature skin aging is generally an issue for extreme athletes only. That said, the use of lots of sunscreen and a great moisturizer can go a long way toward counteracting the relentless pull of gravity.

This story previously appeared on RealSimple.com.

Aging by the Decades

Talk to your dermatologist for good advice about keeping your skin healthy through the years.

30s: You may start to see spotty pigmentation and fine lines near your eyes where the skin creases most readily, particularly if you're a sun worshiper. Try an eye cream to even out and smooth skin.

40s: Tiny lines begin to appear around the eyes and the forehead, and crow's-feet start to set in. Broken capillaries or brown spots may show up on the cheeks, and pores can look larger, says Dr. Lisa Donofrio, an associate clinical professor of dermatology at Yale University School of Medicine.

50s: Deep forehead furrows become more prominent, and smile lines crop up. Blotchiness on the cheeks worsens. And menopause can make women's skin drier.

60s: The face loses volume and sags. Fine lines become more pronounced, and those who have had lots of sun exposure typically get brown patches that may be raised and rough, notes dermatologist Leslie Baumann.

Why We're Spending So Much on Botox

Some types of plastic surgery are on the rise because of the sluggish economy.

✳ BY MARTHA C. WHITE

N 2012, AMERICANS SPENT MORE THAN IN the previous year on products and procedures to make our faces look better. The reason? Well, it may seem counterintuitive, but experts say the lackluster economy is part of the impetus for our collective vanity.

The American Society of Plastic Surgeons (ASPS) says that although total cosmetic surgeries fell by 2% in 2012, the number of what it calls "minimally invasive" procedures rose by 6%. The most popular of these were injections of Botox and Dysport (brand names for botulinum toxin), followed by soft-tissue-filler injections, chemical peels, laser hair removal and microdermabrasion.

The previous year, the number of both cosmetic surgical and minimally invasive procedures rose, although the uptick in surgical procedures was the smaller increase of the two.

"Facial rejuvenation procedures, both surgical and

The popularity of arm-lifts has surged, growing 4,378% over the past decade.

There were 14.6 million cosmetic procedures in 2012, up 5% from 2011.

The top 5 less invasive procedures? In order of popularity:
Botox, soft-tissue fillers, chemical peels, laser hair removal and microdermabrasion.

SOURCE: AMERICAN SOCIETY OF PLASTIC SURGEONS, 2012 STATISTICS

minimally invasive, experienced the most growth in 2012," an ASPS press release states. That includes a record-high 6.1 million botulinum-toxin injections to freeze our frown lines and crows' feet. And although the overall number of surgeries fell, the ASPS says demand for face-lifts and eyelid surgeries rose 6% and 4%, respectively.

The so-called lipstick effect is something consumer psychologists trot out as soon as the economy heads south. The theory goes that we cut back on big-ticket spending but buy ourselves little indulgences as consolation prizes. Instead of springing for a new suit, maybe we'll pick up that designer's cologne. Instead of a pair of pricey pumps, we'll settle for the aforementioned lipstick. Or in this case, we'll go for Botox instead of a pricier nose job or tummy tuck. Maybe we can start calling it the "injection effect."

Wealthier Americans seem more willing to keep buying in order to look good. A recent survey by Unity Marketing, which examines the spending patterns of affluent Americans, found that the rich are becoming more cautious and keeping those platinum cards in their wallets, but company president Pam Danziger says a few spending categories are outliers. For instance, expenditures on beauty services increased 26.5% in the last quarter of 2012, "one of the top growth categories" for that quarter, Danziger says in a report accompanying the survey. "Luxury consumers spent more on spa/salon beauty services in the fourth quarter, showing they are still willing to invest to keep up appearances."

The trend can be seen at the makeup counter, too. In 2012, consumers spent 10% more on department-store-brand skin-care products and 7% more on department-store makeup, according to market-research firm NPD Group. "We have a clientele that's engaged and wants to buy," says Karen Grant, global beauty-industry analyst at NPD.

All this stuff that we use to make ourselves look good supposedly has a by-product effect of making us feel good, too. "The reason we hear most is, 'I'll continue to buy beauty because it makes me feel better about myself,' " says Grant. "This driver is more pronounced in the prestige category. There's a more emotional reason than purely logical."

The willingness to spend is even more pronounced at the upper edge of the price spectrum, Grant says—but it's not just wealthy Americans dropping big bucks on eye cream and eau de toilette. People are buying these little luxuries whether they can easily afford them or not, she says. "They'll find the means at the expense of other things. It's very much an investment. In some cases, you're talking about $300 gift sets and things like that."

There's an indication that for some of us, this spending could be an investment in our careers. One study reportedly found that a worker who ranked among the bottom one-seventh in looks (in the judgment of random observers) earned 10% to 15% less per year than someone whose attractiveness was considered to be in the top third. That amounted to a typical lifetime difference of about $230,000.

In an earlier, equally depressing paper, the Federal Reserve Bank of St. Louis cited research by Daniel Hamermesh, University of Texas at Austin economics professor and author of *Beauty Pays*, that an unattractive worker's "plainness penalty" is 9%, and that there is a 5% "beauty premium" benefiting the pretty and handsome at work.

In 2010, the *Chicago Tribune* noted that older workers aren't just relying on their experience to get ahead in the workplace: they're increasingly trying to turn back the clock with procedures like eye lifts, teeth whitening and hair-loss treatments. "While most older job-seekers know the importance of keeping their skills current, some are applying that same advice to their faces," the article stated.

Some recent research also suggests that increased beauty spending is an investment in our romantic futures, particularly for women. In a paper published in 2012, Sarah E. Hill, associate professor of social psychology at Texas Christian University, wrote that recessions make women work harder to try to attract men, prompting a surge in spending on beauty and cosmetic products and services.

The basic idea is that recessions create a scarcity of financially stable men, so women compete more aggressively for a smaller number of successful, well-to-do bachelors. In experiments, Hill found that female subjects conditioned to think about a bad economy were more likely to display a preference for buying items that could enhance their physical appearance. "Consumers may prioritize beauty," she wrote, "during times of economic turmoil."

12 Things That Can Ruin Your Smile

What you're eating and drinking, along with other habits, could put your pearly whites at risk.

✳ BY KRISTIN KOCH

Your smile is one of your best assets, so you want to keep it sparkling. But even if you brush, use white strips and visit your dentist twice a year, it may not be enough. Here are some factors that can wreak havoc on your teeth and gums.

Sports Drinks In the past decade, sports beverages have become increasingly popular, but they aren't great for your teeth. "Scientific research has found that the pH levels in many sports drinks could lead to tooth erosion due to their high concentration of acidic components, which could wear away at the tooth's enamel," says Dr. David Halpern, spokesman for the Academy of General Dentistry (AGD). Additionally, these drinks are often high in sugars that act as "food" for acid-producing bacteria, which then sneak into the cracks and crevices in your teeth, causing cavities and decay.

Bottled Water Tap water often contains fluoride; about 60% of people in the U.S. have fluoride in their water supply. However, most bottled waters contain less fluoride than is recommended for good oral health (it will be listed as an ingredient on the label if it is an additive). "Fluoride makes the entire tooth structure more resistant to decay and promotes remineralization, which aids in repairing early decay before damage is even visible," explains Jersey City, N.J., dentist Dr. Charles Perle. "Studies have confirmed that the most effective source of fluoride is water fluoridation."

Teeth Grinding Also known as bruxism, this condition can affect your jaw, cause pain and even change the appearance of your face. "People who have otherwise healthy teeth and gums can clench so often and so hard that over time, they wear away their tooth's enamel, causing chipping and sensitivity," says Halpern. Stress and anger can increase nighttime teeth grinding. "Find-

ing ways to alleviate these feelings can help, but it's also important to see your dentist, who can recommend solutions like a custom night guard," advises Perle.

Wine Oenophiles, beware: regular wine consumption can harm tooth enamel. According to Halpern, the acidity can dissolve tooth structure, and both red and white wine can increase dental staining. Still, you don't have to give up your regular glass of vino to save your smile. "Enamel erosion develops when wine drinkers swish the wine, keeping it in constant contact with the enamel," says Perle. "So instead, take small sips and rinse with water when you're done drinking."

Dry Mouth A dry mouth isn't just unpleasant, it's bad for your teeth. Saliva washes away cavity-causing bacteria and neutralizes harmful acids. "Without saliva, you would lose your teeth much faster—it helps prevent tooth decay and other oral-health problems," says Dr. Gigi Meinecke, a dentist in Potomac, Md. Drink lots of water, chew sugarless gum, use a fluoride toothpaste or rinse, and consider over-the-counter artificial saliva substitutes. See your doctor if it's a frequent problem.

Citrus Fruits "Although lemons, grapefruits and citrus juices don't directly cause cavities, they contain acids, which cause erosion of the tooth enamel, weakening the tooth and making it prone to decay," says Meinecke. Rinsing your mouth with water or chewing sugar-free gum can help. Xylitol, a natural sweetener added to sugar-free gum, mints and toothpastes, can inhibit oral bacteria that cause cavities. Meinecke suggests that you chew at least two pieces of xylitol-containing gum per day if you're at high risk for developing cavities.

Dieting Restrictive diets and poor eating habits can deprive you of the vitamins and nutrients necessary for a beautiful smile. It's especially important to get enough folate, B vitamins, protein, calcium and vitamin C; all are considered essential for healthy teeth and gums. And, says Halpern, "poor nutrition can affect your entire immune system, increasing your susceptibility to many disorders and infections, including periodontal disease."

Aging As you age, you're more susceptible to decay near old fillings or root surfaces unprotected by receding gums. But there's no reason you can't keep your teeth. Oral disease—not aging per se—is the danger. Bumping up your fluoride protection is key. And if you have arthritis, there are dental products that can make brushing and flossing less painful. "Seniors who brush regularly with fluoride toothpaste or use fluoride rinses or gels regularly have fewer cavities," says Meinecke.

Birth-Control Pills Because they mimic pregnancy—when changes in hormone levels can cause gum inflammation—oral contraceptives can also lead to inflammation and infections, including gingivitis. Additionally, some studies have shown that women who use birth-control pills may have more trouble healing after tooth extractions and are twice as likely to develop painful dry sockets where the tooth used to be. If you use birth-control pills, discuss their effect with your dentist before major procedures.

Not Flossing Although many of us are much more diligent about brushing than flossing, they are equally important. "Flossing every day is one of the best things you can do to take care of your teeth. It's the single most important factor in preventing periodontal disease, which affects more than 50% of adults," says Meinecke. Flossing helps remove plaque and debris that stick to teeth and gums, polishes the tooth's surface, and even helps control bad breath.

Brushing at the Wrong Time We've been taught to brush after every meal, but depending on what you eat or drink, that's not always best. "After consuming high-acid food or drinks like wine, coffee, citrus fruits and soft drinks, rinse with water to neutralize the acids, but wait an hour before reaching for the toothpaste," says Meinecke. "Brushing teeth immediately afterward can cause erosion."

Overzealous Whitening It's not clear whether bleaching erodes tooth enamel, but it can increase sensitivity, especially when done too often. Even at-home treatments should be used with moderation, as some whitening toothpastes and gels contain abrasive ingredients that can exacerbate sensitivity. And as you get older, whitening products can only do so much; after a certain point, using more won't necessarily do anything for your smile.

This article originally appeared on Health.com.

Hair Today, and Tomorrow

Losing your locks can be emotionally devastating. Thankfully, science is devising promising new options to help those of us at the mercy of hair loss.

 BY COLLEEN MORIARTY

HAIR LOSS (OR ALOPECIA, AS it's medically known) provides stand-up comedians with endless material—"I have a feeling dead people get really mad when we complain about losing hair," says Louis C.K.—but for most people, it's no laughing matter. "People's sense of identity is closely tied to their hair," making this a highly emotional and distressing problem, says Dr. Luis Garza, an assistant professor of dermatology at Johns Hopkins School of Medicine and a leader in hair-loss research.

When it comes to hair loss, family genes are often—but not always—to blame. "There are many reasons hair can fall out," says Dr. Mary Lupo, a New Orleans–based dermatologist and a past president of the Women's Dermatologic Society. Luckily for those who are losing, many causes are reversible, she adds. A doctor can help pinpoint the source of the problem and discuss possible treatments, which now run the gamut from topical solutions to high-tech follicle transplants that use an FDA-approved robot.

An Inheritance No One Wants

Heredity is responsible for the most common type of hair loss: androgenetic alopecia, a medical condition that affects about 80 million American men and women, according to the American Academy of Dermatology (AAD). It's caused by excess male hormones: specifically, too much androgen and dihydrotestosterone (DHT).

In men, androgenetic alopecia looks very familiar. It's the classic receding hairline, or the loss of hair on the top of the scalp—what's termed male pattern baldness. A guy has a 50% chance of experiencing this sort of balding by his 50th birthday, according to the AAD. Female pattern hair loss, on the other hand, occurs when hormones get out of whack, as during menopause or when a woman has ovarian cysts, an endocrine condition or a thyroid issue, says the American Hair Loss Association. You can recognize it by a widening part and diffuse, all-over thinning.

While hair loss can be upsetting to both genders, balding women may be particularly hard hit, especially since there's really no female version of a "power buzz" haircut. "When I see women in consultation about their hair loss, they are often emotionally distraught and demoralized," says Dr. Ivan Cohen, a clinical professor of dermatology at Yale. "Men—especially younger men—are often embarrassed by their thinning hair if it makes them look older than their peers."

Root Causes

Even if you escaped a family legacy that includes a balding pate, there's still a small chance you'll fall victim to another type of hair loss. Alopecia areata, an autoimmune disorder, causes hair to suddenly come out in clumps as the body's immune system attacks its own hair follicles on the head, face and body. Currently, there's no cure or even a treatment that works for everyone with the disorder, but most patients do see their

hair regrow, according to the National Alopecia Areata Foundation.

Research led by Angela Christiano of Columbia University Medical Center, who has suffered from alopecia areata herself, has identified eight genes that contribute to this kind of hair loss, including one that looks to be the instigator of the disease. Interestingly, a number of the genes are also associated with other autoimmune diseases, such as rheumatoid arthritis, Type 1 diabetes and celiac disease. Clinical trials are under way to see if medications for these conditions, especially rheumatoid-arthritis drugs, might effectively treat alopecia areata, too.

So let's say you were blessed with hair-friendly genes, but you're still watching too many strands go down the shower drain (it's normal to lose about 50 to 100 hairs every day). These triggers could be to blame:

A POOR DIET: Hair loss has been linked to eating disorders; deficiencies in substances such as iron, protein, certain minerals and the amino acid lysine; and too much vitamin A (which can happen if you overdo supplements). Even too-rapid weight reduction can cause lost locks months later, though hair usually comes back on its own, says the AAD.

A new computer-assisted robot harvests thousands of healthy follicles that are then transplanted to patches where hair has been lost.

ILLNESS: About 30 diseases are known to cause hair loss; high fever, flu, thyroid disease and infections can also have an impact. Excessive shedding can be a side effect of some medications, including birth-control pills, drugs for hypertension (beta blockers), the acne medicine Accutane, blood thinners and antidepressants, says Cohen. And while we know that chemotherapy drugs bring on balding, radiation and surgery can also cause thinning. During an exam, a dermatologist will check for ringworm of the scalp, which causes balding unless treated with oral antifungal medications.

STRESS OR TRAUMA: Extremely stressful situations, like the death of a loved one or a traumatic accident, can trigger chemical changes that disrupt the normal hair-growth cycle, what is medically termed telogen effluvium. Even childbirth can bring on this condition. Normal growth, though, typically returns within a few months.

HAIRSTYLING: Harsh styling—think frequent bleaching, aggressive brushing, rough relaxers or bad perms, as well as wearing too-tight ponytails, braids or cornrows—can lead to localized hair loss rather than diffuse thinning. This type of hair loss "can be irreversible if the damage is continued," cautions Lupo.

Get-Growing Solutions

If heredity is to blame for your baldness and you're a guy, there are treatments: minoxidil 5%, a topical scalp treatment (found in Rogaine); ultraviolet-light therapy to stimulate hair growth; and Propecia, a prescription tablet with the active ingredient finasteride—although it has side effects that can include loss of erection and decreased libido in men. Propecia is not FDA-approved for women, but a lower dosage of minoxidil (2%) and laser therapy are options. Doctors treat alopecia areata with a combination of steroid injections, topical application of minoxidil, corticosteroids and immunotherapy, along with ultraviolet-light therapy.

What's really promising, though, is the work now going on in labs around the world: researchers are using platelet-derived growth factors to induce hair growth, says Dr. Neil Sadick, a dermatologist in New York. That means substances are taken from your own blood and injected into your scalp to stimulate cell growth in the follicles—a procedure called platelet-rich plasma (PRP) therapy. "Excellent results have been achieved with PRP and documented to increase hair thickness," says Sadick.

Hair-replacement techniques have also come a long way since hair plugs, thanks to the FDA-approved ARTAS Robotic System. It uses a computer-assisted robot to harvest hundreds—even thousands—of individual hair-producing follicles from the "donor" area of the scalp (places on the head where there's no balding), leaving just tiny circular incisions that heal quickly, Cohen explains. The healthy follicles are then transplanted to patches where hair has been lost. "Recovery is shorter, with less postoperative discomfort than earlier transplant methods—and there is no scar," he says. The price for a robot-assisted restoration ranges from a few thousand dollars for a small procedure to upward of $10,000 for extensive baldness. "If the balding area is relatively small, then the results of the robotic transplant will look as full as it looked before," Cohen says.

Dr. George Cotsarelis of the University of Pennsylvania's Perelman School of Medicine and Garza of Johns Hopkins have discovered that elevated levels of PGD_2 (prostaglandin D_2) hold back hair growth in men with androgenetic alopecia. PGD_2 acts through a receptor on the cell called DP-2. "Pharmaceutical companies have found that this receptor is important in some allergic diseases, so they are already developing drugs to target DP-2," says Garza. Cotsarelis's lab will be testing different drugs to see if they improve hair growth. "We hope that if our work leads to successful therapies," says Garza, "we might help those who suffer from the psychological effects of alopecia." Which are deep-rooted indeed.

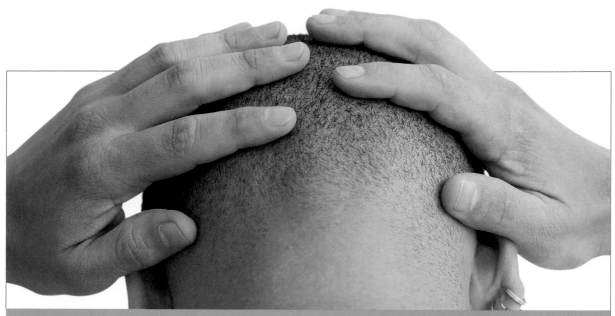

Shave It Off!

How bald guys can look more manly and dominant.

BY ALEXANDRA SIFFERLIN

FORGET HAIR TRANSPLANTS, PILLS AND regrowth serums. A University of Pennsylvania researcher has a better alternative for balding men: shave it all off.

In three experiments, Albert Mannes, a lecturer at Penn's Wharton School and a balding man himself, found that guys with shaved heads are perceived by others as not only more manly and dominant than other men, but also taller, stronger and having greater potential as leaders.

In the first experiment, nearly 60 participants looked at a series of photos of men who were similar in age and attire. The difference was that some had shaved heads, while others had full manes. The participants rated each man in terms of how powerful, influential and authoritative he looked. When the results were averaged, shaved men topped the ratings.

Next, Mannes showed participants images of four men. Each was shown twice, once with hair and once without. Not only were the men perceived as more dominant when they were shown digitally balded, but they were also viewed as nearly an inch taller and 13% stronger.

Finally, the participants were given verbal and written descriptions of the men. Some were described as having thick hair, others as having shaved heads. Once again, the participants rated those with shaved heads highest for masculinity, strength, dominance and leadership potential.

"I was surprised that perceptions of dominance and masculinity extended to concrete physical char-

acteristics such as height and strength," says Mannes, whose study was published in the journal *Social Psychological and Personality Science*.

Mannes theorizes that it's the boldness of the act of head-shaving that feeds into the perception of dominance. He has found that men with thinning hair—those who are presumably just resigning themselves to their own baldness—were rated as less dominant than those who took the initiative to shave their heads altogether.

Still, that doesn't mean everyone should be grabbing the razor. Mannes's research also revealed that guys with shaved heads were considered less attractive and older-looking than those with thick heads of hair—and attractiveness is also correlated with perceptions of dominance. "So whatever a man gains in dominance directly by shaving his full head of hair will be offset to some degree by his diminished attractiveness," he says.

For those with thinning hair, however, the benefits offset the downside. "The shaved look is more attractive than the visibly balding look," says Mannes. "So men suffering natural hair loss may enhance both their dominance and attractiveness by shaving."

Mannes says his findings should help legions of balding guys feel better about themselves and stop being self-conscious about their shiny pates. Or at the very least, they can save themselves the financial cost of trying to reverse their hair loss: "These men might better improve their well-being by finishing what Mother Nature has started."

Going Strong

The Ideal Exercises
Four Mistakes You're Making at the Gym
Five Workouts to Mix It Up
The Perfect Playlist
Goal Power

NEXT TO THE FOODS WE PUT INTO OUR MOUTHS, there's simply no better prescription for youthfulness—inside and out—than regular physical activity. But which kind, and for how long? Science is helping to answer both questions. You'll probably be relieved to hear that research shows you have to work out a lot less than you think, and that there's no "perfect" exercise either (an ideal regimen combines three key elements). Looking for motivation? Two words: music and variety. Scientists have actually studied how to put together the perfect playlist, and fun, innovative workouts (upside-down yoga, anyone?) go a long way to stave off boredom, the biggest enemy of regular exercise. In this chapter you'll also get help setting and achieving your goals, with steps that will take you from "I know I should do this" to "I'm actually doing it!"

The Ideal Exercises

There's no such thing as one perfect workout. But if you add in elements of three specific types of training, you'll get pretty close.

✳ BY MYATT MURPHY

GETTING A STRONGER HEART, healthier muscles and less excess body fat isn't simply a fitness finish line most people wish they could cross. It's a race many of us never bother to enter.

According to the Centers for Disease Control and Prevention (CDC), approximately two-thirds of all adults over the age of 20 in the U.S. are overweight, with more than half of that group considered obese. It's a staggering number that simply doesn't measure up, especially when the benefits of staying fit go way beyond liking what you see in the mirror.

For decades, regular exercise has been shown to accomplish everything from decreasing your risk of cancer, diabetes, hypertension and high cholesterol to turbo-boosting your mood, energy level and sex drive. New research published in the *American Journal of Preventive Medicine* has discovered that doing some form of exercise for at least two and a half hours a week could extend your life by as much as five years.

With those perks waiting for those willing to sweat, why would anyone decide not to exercise? It's fairly simple. For most of us, it's not that we choose not to exercise—it's just that we have no idea what to choose when it comes to exercise.

At every turn there seems to be some book, DVD or fitness product offering a "new" secret to getting in shape. But the real secret is that there isn't any secret at all. The key to feeling healthier, getting stronger, staying injury-free and keeping extra weight at bay begins with understanding the basic components of fitness and the ideal way to incorporate each into your life. That's it.

If you're ready to join the one-third of Americans who know that the easiest way to get a handle on their waistline is by not wasting their time, here are the three cornerstones that make up the ideal fitness regimen—and how to perform all three the right way every single time.

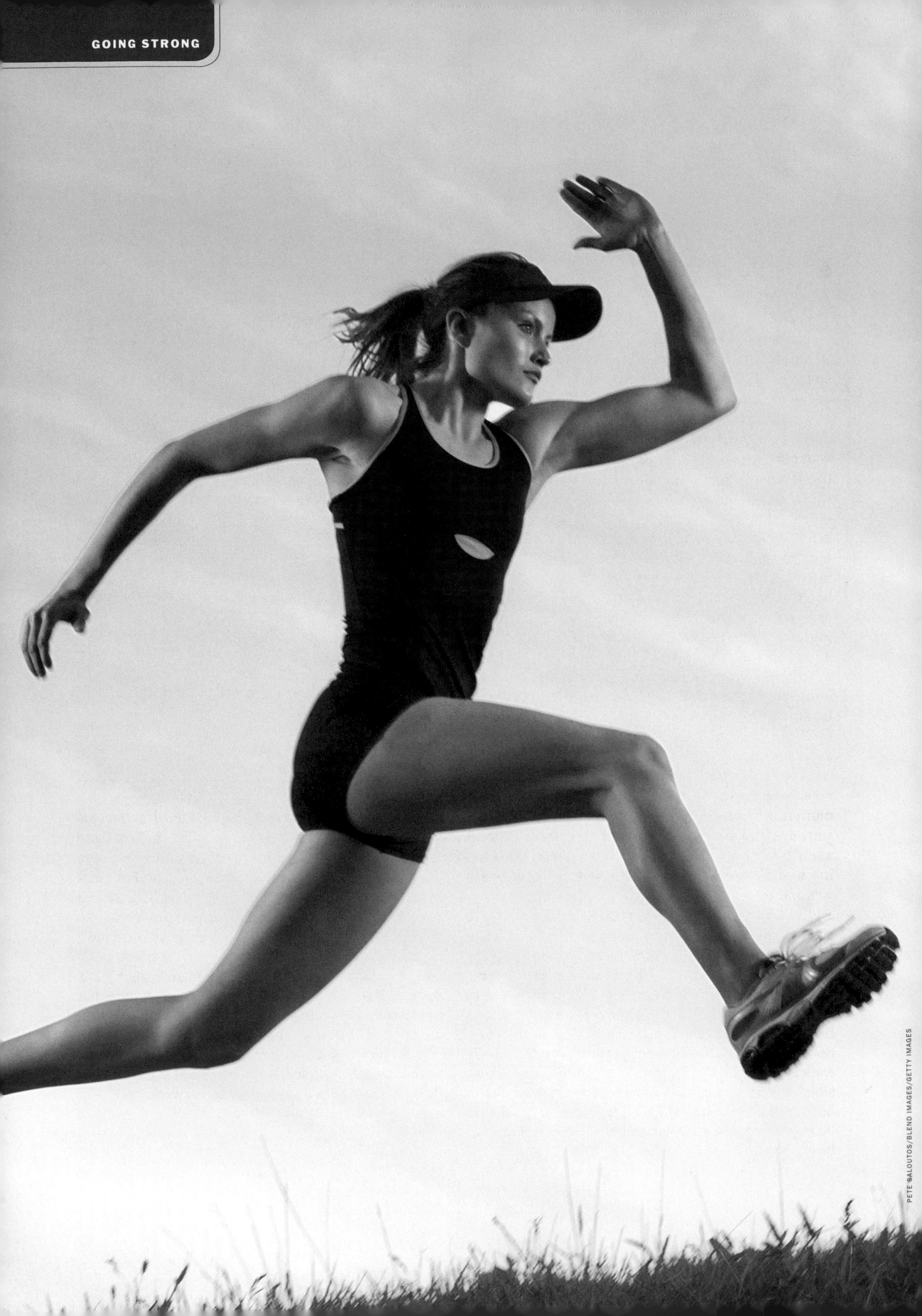

PETE SALOUTOS/BLEND IMAGES/GETTY IMAGES

1. Cardiovascular Exercise

THE FIRST—AND WHAT MANY EXPERTS believe is the most vital—component of keeping fit is cardiovascular exercise (or "cardio"), which is any activity that increases your heart rate.

That's right, we said any activity. Despite the popularity and effectiveness of certain types of cardio such as running and cycling, just three things matter to your heart: frequency, intensity and time. It doesn't care what activity makes your pulse quicken, whether you're playing hopscotch with your kids, pushing your lawn mower or returning serves on the tennis court, it all counts.

What Science Says If you're easygoing, the CDC recommends a minimum of 150 minutes of "moderate-intensity" cardio activity every week. For even greater health benefits—and if weight loss is your ultimate goal—increase that to 300 minutes. If that sounds daunting, just remember that you're not expected to do it all at once. "Although most people prefer doing cardio in 20- to 30-minute bursts, it still counts if you feel like doing less," says Richard T. Cotton, national director of certification for the American College of Sports Medicine (ACSM). "Research supports that sessions of as little as 10 minutes are still as effective."

So what's "moderate intensity"? Any tempo that raises your pulse to between 50% and 70% of your maximum heart rate (MHR), which is estimated by taking your age and subtracting it from 220. To measure it another way, if you can talk as you exercise but singing would be too difficult, you're somewhere between 50% and 70%.

If you're time-starved, the CDC says it's possible to achieve the same results by doing only 75 minutes of cardio (and up to 150 for greater health benefits). The not-so-good news is that all those minutes have to come from high-intensity activity that raises your pulse to around 85% of your MHR. To figure out what that feels like, Cotton recommends using the OMNI Scale of Perceived Exertion, which requires rating how hard you're exercising on a scale of 1 to 10 (1 being "as little as possible" and 10 being "as hard as possible"). "In order to be sure you're working out at high intensity, you should be exercising at a pace somewhere between 8 and 9," he explains.

How to Pull It Off First, see a doctor and get a health screening to determine if you're ready to start a program, advises Barbara Bushman, professor of kinesiology at Missouri State University and editor of the ACSM's *Complete Guide to Fitness & Health.* "Every exercise program needs to start where individuals are rather than where they want to be," she says. Once you have the green light, choose an activity. If losing body fat is your goal, we won't lie: running, hill sprinting, swimming, cross-country skiing and skipping rope will burn more calories than activities such as walking or light cycling. Which is best? That's easy—whichever one you're least likely to quit.

Once you've decided on an activity, easygoing exercisers can follow a "steady-state" approach—an even, unvarying moderate pace for the entire workout. If you're time-starved, High-Intensity Interval Training (HIIT) requires that you combine short bursts of high-intensity exercise with longer bursts at low intensity. So instead of running at the same boring pace, to use HIIT you'd alternate between sprinting for 15, 30 or 60 seconds and lightly jogging at low intensity for roughly twice as long (either 30, 60 or 120 seconds), for the entire workout. HIIT has been shown to improve stamina and speed, force your body to use more fat as

---

Just three 30-minute sessions of Sprint Interval Training were as effective as performing

5 hours

of steady-state exercise.

---

fuel, and even help build and preserve more lean muscle.

There's also Sprint Interval Training (SIT), a more aggressive approach that involves alternating between 30-second "all-out" sprints and four and a half minutes of very-low-intensity exercise. According to research published in *The Journal of Physiology,* just three 30-minute sessions of SIT were as effective as five hours' worth of steady-state exercise.

The downside: because of the demands that HIIT and SIT place on your body, doing too much of either can actually slow your progress, or worse, leave you more susceptible to injury. That's why most experts prefer that beginners stick with steady-state cardio and recommend that intermediate to advanced exercisers use HIIT no more than three or four days a week.

2. Resistance Training

I**F THE THRILL OF LIFTING WEIGHTS ISN'T YOUR THING, YOU'RE NOT** alone. As popular as some people may believe strength training is, according to a recent study published in the *Journal of Strength and Conditioning Research*, only 9% of Americans bother to do it. The problem is that even if you couldn't care less about getting stronger, having more lean muscle plays a more significant part in improving your overall health than you might realize.

For example, that same research found that men and women who lifted weights were less likely to have metabolic syndrome, a cluster of five risk factors—ranging from a large waist to high blood pressure—that have been shown to increase your risk of developing heart disease and diabetes. Building more muscle through resistance training also has a key role in ensuring stronger bones, ligaments, and tendons; improving your balance; and assisting in weight control, since individuals with more lean muscle have a higher resting metabolic rate—the amount of calories your body burns all day long.

That's why resistance training is so crucial for older adults. "After sedentary men and women turn 40, they begin to lose on average between 5% and 10% of their muscle mass every decade," says Darryn S. Willoughby, associate professor of exercise/nutritional biochemistry and molecular physiology at Baylor University. "Since muscle tissue uses more calories to meet its energy needs, losing it lowers your metabolic rate and you begin to use fewer calories."

That decrease in metabolic rate, says Willoughby, is a primary factor in why we begin to accumulate body fat as we age. Maintaining and increasing your muscle mass through some form of resistance training can help prevent that from ever happening and, according to the CDC, can raise your metabolism by as much as 15%.

What Science Says In 2011, the ACSM released the most up-to-date recommendations on the quantity and quality of exercise for adults. According to the organization, the average adult should train each major muscle group two or three days each week, waiting at least 48 hours between resistance-training sessions.

Why that long? Your muscles need 48 to 72 hours of recovery time to heal and

After sedentary men and women turn

40,

they begin to lose on average

5%-10%

of their muscle mass every decade.

return to their usual level of performance. Doing too much resistance training without proper rest can cause a series of negative effects, including a decline in muscle strength, a greater likelihood of suffering from an injury or becoming sick, and a rise in cortisol, a hormone responsible for breaking down muscle and storing fat.

How to Pull It Off Where many people get confused is the "how" when it comes to resistance training, which is basically any exercise that forces your muscles to contract under some form of resistance. That resistance can come from almost anyplace: dumbbells, barbells, exercise machines, stretch cords, the old cinder block in your garage, or even your own body weight—when doing pull-ups, push-ups or squats, for example.

How many reps (a single motion of an exercise) and sets (a group of consecutive reps performed without resting) you should do of any given exercise depends on your training goals. The ACSM's latest research found that performing each exercise for two to four sets helps adults improve muscle strength, power and endurance, although doing just one set of any exercise can also be effective, especially for older and novice exercisers.

When it comes to reps, the data is fairly straightforward, according to the ACSM's new guidelines: if you're looking to improve muscular strength and power, choose a weight that allows you to perform 8 to 12 repetitions of each exercise. If you're middle-aged or older, however, or a beginner to resistance training, the best range is 10 to 15 repetitions. Finally, if you participate in other activities and would benefit more from having muscular endurance, stick with 15 to 20 repetitions per set. No matter which rep range works best for you, rest for two to three minutes between sets.

When you're choosing the type of resistance or weight, there's an option for every exerciser. Resistance bands are light, easy to store, versatile and inexpensive, and they travel well. If you prefer to lift actual weights, then barbells, dumbbells and kettlebells are also versatile and can be purchased in a variety of weight loads. If you're not sure of proper form, or how to use bands or free weights, consider working with a personal trainer for a few sessions until you feel comfortable continuing on your own.

3. Flexibility

ALTHOUGH IT OFTEN TAKES A BACK-seat to cardio and resistance training in the minds of most people, flexibility is considered the third key component of fitness with good reason. Too many people dismiss flexibility because they assume that it's something only athletes or extremely active people can benefit from, or that they'll need to work their way into some crazy yoga pose. But being able to move your joints through their complete range of motion—the definition of flexibility, according to the ACSM—is critical for improving your overall performance and reducing your risk of injury while performing any activity, from the strenuous (lifting a heavy box) to the mundane (bending down to tie your shoes).

"Staying limber alleviates stress, improves your co-ordination and balance, and can protect you from postural issues, particularly if you're already doing regular cardio and resistance training," says Dr. Nicholas Di-Nubile, chief medical officer of the American Council on Exercise and author of the FrameWork Active for Life book series. "That's because both cardio and resistance training can tighten certain ligaments and muscles, particularly within the front of your shoulders, lower back, hamstrings and calves."

What Science Says Unfortunately, flexibility naturally decreases as you age, which is why experts recommend that adults do some type of flexibility program for a minimum of two or three days a week. Unlike cardio and resistance training, however, flexibility is something you can work on every day if you do it properly. No matter how often you decide to do it, your routine should include at least one stretch that targets each major muscle group—especially your chest, shoulders, upper and lower back, abdominals, gluteals (buttocks), quadriceps, hamstrings and calves.

"There has been contradictory research recently that has given stretching a bad name, saying it may not help prevent injuries and could even reduce performance when performed immediately before [exercise]," says DiNubile. For example, research presented in 2011 at the American Academy of Orthopaedic Surgeons found that stretching for three to five minutes before running neither prevents nor causes injuries. Other studies have found that stretching before resistance training may decrease your strength as you exercise.

According to DiNubile, this new research mainly focuses on the performance of younger, advanced athletes,

and these studies may actually be doing older adults a disservice. "Most active baby boomers are tight as a drum prior to activity," warns DiNubile, who emphasizes that warming up your muscles and doing some pre-stretching before any activity is still a safe bet.

How to Pull It Off Experts will tell you that to improve flexibility, there are four types of stretches you can use to loosen up, three of which—ballistic, dynamic and proprioceptive neuromuscular facilitation (PNF)—are generally not recommended for the average person. Even though these three can be remarkably effective, they require certain techniques, such as performing quick movements or contracting your muscles as you stretch, that can increase your risk of injury. So the last type—static stretching, which involves slowly getting into a stretch, then holding it for a certain period—is the recommended option for the average person.

If stretching feels too boring, DiNubile recommends trying yoga. "Yoga offers a few benefits that go beyond flexibility," he says, a notion that science supports. Researchers now believe yoga may be effective at treating stress-related psychological and medical conditions in-

---

Stretch only to the point
of tightness or slight discomfort
(never pain!), hold for

10 to 30

seconds, then repeat each
stretch two to four times.

---

cluding depression, anxiety and even heart disease.

Which stretches you choose are entirely up to you, but what experts highlight when it comes to static stretching is following the rules. First, stretch only after warming up your muscles for at least five to 10 minutes to make them more pliable. "I tell patients to break a light sweat prior to stretching," says DiNubile. That can mean doing your stretches immediately after your cardio or resistance-training workout, after a quick full-body warm-up (such as running in place or doing some jumping jacks), or after a warm shower or bath.

As you settle into each position, stretch only to the point of tightness or slight discomfort (never pain!), hold for 10 to 30 seconds, then repeat each stretch two to four times. DiNubile's final rule of thumb: don't bounce to try to get a greater range of motion. "Flexibility isn't a competition," he says. "In other words, to see the greatest results, don't look at what everyone else can do. Focus on improving your own flexibility a little bit more every day instead."

Four Mistakes You're Making at the Gym

Fitness experts help tweak your workout to make it more effective.

BY ALEXANDRA SIFFERLIN

ILLUSTRATION BY PETER ARKLE

1. Going Overboard on Cardio Machines Getting regular aerobic exercise does wonders for your health, decreasing the risk of obesity and diabetes, strengthening the cardiovascular system and perhaps even staving off Alzheimer's. The problem is that many people aren't getting the maximum benefit from their cardio workouts because they're either using the machines wrong or failing to pace their exercise correctly.

If you're using an elliptical machine, for instance, pay attention to where you set the resistance. "Hunching over or using a death grip on the machine handrail because your incline or resistance is too high for you cheats your body and can throw off your alignment, jarring your spine, shoulders and elbows," says Scott Danberg, director of fitness at the Pritikin Longevity Center in Miami. He suggests challenging yourself enough to make it a tough workout but not so much that you can't have a natural gait with a light grip while you're using the equipment—this goes for any cardio machine.

And try mixing up your routine. Sara Haley, a Reebok Global master trainer and independent fitness consultant, recommends that diehard treadmill joggers add high-intensity machines like the rower or the Jacob's Ladder to their regimen. These machines make cardio more efficient by working more muscle groups and burning more calories.

2. Incorrect Weight-Lifting Many people make judgment errors when lifting weights. Men are more likely to choose ones that are too heavy for them, while women tend to be fearful of bulking up and go for weights that are too light. A recent study found you don't need heavy weights to gain muscle; lighter ones can be just as effective if used correctly. Danberg recommends choosing a weight that you can lift 30 times to start but afterward can lift only 15 times more. "You want to get to your repetition goal but be able to put down the weight and think, 'What's next?'" he says. "You don't want to feel exhausted to the point of 'Oh, God, what did I do?' This will keep you injury-free, but you will still feel the burn."

When it comes down to it, it's all about maintaining correct form to get the most out of resistance training. "When you have incorrect form, chances are you are going to jeopardize your balance," says Danberg. "You work your body harder than it actually has to work." Quick tips for correct form: keep a strong upper back, with your chin and chest up, and tighten your core. And there's nothing wrong with taking it slow, adds Haley. A lot of exercises become more challenging at a more measured pace.

3. Failing to Focus on Your Core "When people are not paying attention to their core at the gym, you can see it," says Danberg. "They stand there as if they are at their kitchen counter." Both Danberg and Haley say the body's core is the basis of all human movement, so strengthening it improves all other physical activities. "Many exercisers only place a focus on contracting the abdominal muscles when doing specific ab exercises such as sit-ups or when using an ab machine," says Danberg. "While this is good, contracting this area during exercise movements such as bench presses, back rows and leg presses will allow better stability during the movement and less risk of injury."

The core refers not only to your abdominal muscles but to the entire area from your chest down to your hips. Danberg tells his clients to always work out in "sport-ready position": standing in a posture you could quickly move from if, say, someone threw a ball your way. "Slightly bend your knees and hold in your abs," he says. "Your posture will immediately improve, and you will feel your whole body working."

4. Ignoring Unseen Muscle Groups People go to great lengths to achieve bulging biceps or well-defined abs, but they often forget about smaller or less visible muscle groups, like those around the joints. "We tend to only target our biggest muscles, which is important because we need those to move," says Danberg. "But we also need to focus on our stabilizing muscles, those smaller muscles around the hips and shoulders. By working those, form improves, and you can actually do more during your workout."

Haley advises people to pay attention to invisible inner muscles, namely those that help control the flow of urine. Kegel exercises—contracting and relaxing the muscles of the pelvic floor—work these deep muscles in both men and women. "These muscles are known to help women have an easier childbirth, and are believed to help in sexual function for men and to help combat incontinence in both genders," she says.

101

Five Workouts to Mix It Up

The best way to beat boredom? Add variety to your effort. We've put these regimens to the test and found all effective—and even fun.

✳ BY ALEXANDRA SIFFERLIN

F THERE'S ONE THING THAT'S THE ENEMY of motivation, it has to be boredom. Never is this truer than when it comes to exercise. In fact, some research shows that half of those who start a workout program drop out within the first six months. After all, few of us feel like born athletes, and even the pros have days when they just don't feel like exercising. Which is why finding inno-vative regimens, classes, instructors and equipment can go a long way toward getting you going and keeping you moving.

Trying new workouts may also be the best way to keep seeing results; research shows that mixing up your training—varying that well-trod jogging path around the park with bouts of swimming, Pilates or hiking, for example—will help work your muscles and cardiovascular system harder and prevent the accli-mation that, over time, means you stop seeing results. (Nobody likes a plateau, right?)

So if you're looking to feel more fit, toned and strong, any of these five challenging workouts will fit the bill. They vary from demanding routines that use only body weight (think push-ups, pull-ups, lunges and crunches) to ballet-inspired regimens and a version of yoga that makes it a whole lot easier to go upside down. I put all five to the test, and each one offers something different and interesting to keep you having fun and seeing results—even while you're working out.

Upside-Down Yoga

What It Is Aerial vinyasa, also referred to as suspen-sion or antigravity yoga, is one of many yoga-fusion techniques. It blends music and circus acrobatics with classic yoga moves, incorporating a sturdy silk sling suspended from the ceiling. The sling supports you as you hold challenging positions or move from one position to another, and you also hang from it (or in my case, get tangled up in it) during the upside-down inversions.

What the Experts Say "The practice works your whole body and allows you to do moves without compressing your spine," says instructor Angelina Borodiansky at OMFactory in New York. "You will feel it all over, and it can increase your flexibility and overall body circu-lation. With the hammocks, you're going to be using muscles in different ways than you would in a regular vinyasa [flow] class."

What It Feels Like During the class, I found myself dangling in an inverse triangle pose by my feet. Spending time upside down has long been a part of yoga tradition, but handstands require a great deal of upper-body strength. The slings—which hold up to 1,000 pounds—make it easier for less experienced yogis to make the flip. "Everyone can benefit from this," says Borodiansky. "The hammock really brings out the kid in the yogi."

*In aerial vinyasa yoga,
a silk sling provides support as
you hold challenging positions.
You'll enjoy yourself, too.*

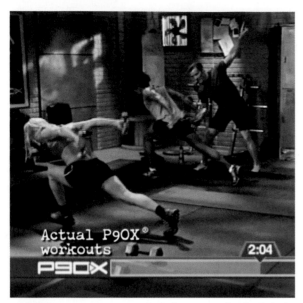

Many of the exercises in P90X involve little equipment besides the body.

Ballet-based exercise classes include various levels of true dance skills.

P90X's Muscle Confusion

What It Is A popular 90-day body-sculpting program that includes 12 workout DVDs, a 100-page fitness guide, a nutrition plan, and a 90-day calendar to track your progress; it's sold more than four million copies since 2005. Workouts incorporate a range of exercises—including chest and back strengthening, plyometrics and yoga—designed to create "muscle confusion," a cross-training regimen that's supposed to keep the body from adapting to the same exercise over time and prevent plateaus.

What the Experts Say "Muscle confusion is a variety of exercises," explains P90X creator Tony Horton. "Typically, a lot of programs focus purely on stationary yoga, Pilates, weight training or cardiovascular; P90X has all of those. The main purpose is to avoid the three things that cause most people to stop exercising: boredom, injuries and plateaus. In P90X, there's variety in fitness with muscle confusion, and the goal is for people to get through the whole 90 days so they have real success and real results."

What It Feels Like P90X is broken up into cycles. "Each month there is a different sequence of exercises," says Horton. The first three weeks are basic exercises, followed by a recovery week. The second month brings a set of more challenging routines, followed by another recovery week. ("Recovery means you are still working hard," just not quite as rigorously as during the first three weeks of the month, explains Horton.) The third and final month is a combination of what you did in the first two months.

Body-Busting Barre

What It Is With long limbs and lean torsos, dancers' bodies are much coveted. Barre is one of the most popular dance-inspired workouts to come center stage, and although it isn't new to the fitness industry, branches inspired by the original practice continue to pop up around the country. The barre method is largely credited to former German dancer Lotte Berk, who started teaching her classes, even to nonballerinas, in the 1970s in New York.

What the Experts Say "Barre can be any number of different techniques. Every studio has their own flavor," says Kaitlin Vandura, co-owner of New York's Pure Barre. "We use really tiny, isometric movements. I would liken it to the concept of using light weights and doing a lot of reps and really burning out the muscle."

What It Feels Like Most barre classes include the use of—you guessed it—a ballet barre where students work through a series of skills that tone arms, torso, feet and legs. The goal is to strengthen, lengthen and sculpt the body with exercises that infuse Pilates and weight training into classical ballet techniques. In a typical workout you might find yourself doing several reps of small, pulsing muscle contractions; some studios mix barre work with exercises on suspension bands, using small weights and cardio work. Afterward, expect to feel some shakiness. "It means you worked the muscle to fatigue," says Vandura, "and the science behind that is the muscle then builds itself back up stronger and leaner." Start with a beginners' class to learn the correct form.

Training with serious weights is just part of the regimen at CrossFit.

One solution for those who say they don't have the time to work out.

Strength-Building Boot Camps

What It Is Weight-training classes and boot camps that target anyone who wants to get strong, not just lose weight. These workouts include circuit rounds of running, calisthenics, free weights and obstacle courses. CrossFit started in the mid-1970s in Santa Cruz, Calif., attracting police academies and Marines who were looking for heavy-duty boot camps with weights.

What the Experts Say Like many popular boot camps, Warrior Fitness Boot Camp is run by two former Marines and is based on their own military training. Hard-core strength challenges are what Alex Fell, co-owner of the New York City Warrior Fitness Boot Camp, says keeps clients returning. "[They] see tremendous results and they get addicted to the workout," he says. "They've never done anything this intense, and they've never thought they could do anything like this before. They've never climbed over walls and up ropes. Every time they come in, they overcome a new obstacle."

What It Feels Like At Warrior Fitness Boot Camp I found myself panting, soaked in sweat, and staring down a six-foot wall. In what can only be explained as a miracle, I forced myself over the wall with plenty of slipping and heaving. I also enrolled in CrossFit NYC's six-class workshop, where I worked at perfecting push-ups, squats and killer pull-ups, followed by circuits of strength training to exhaustion. I wasn't totally sold on heavy weights, but focusing on getting fit and being obsessed with burning calories does make working out much more fun.

Tabata: High-Intensity Training

What It Is Tabata Protocol is a scientifically studied method to boost endurance in only 25 minutes or so. It's based on studies of high-intensity intermittent training by Japanese researcher Izumi Tabata, in which you train at your absolute maximum level for short bursts of 20 seconds, then recover for 10 seconds, and repeat the cycle numerous times. If followed correctly, Tabata should fatigue even the fittest gym bunnies.

What the Experts Say In the Tabata class that I sampled in New York, instructor Amanda Young led us in rounds of ferocious step-ups, lunges and push-ups. Twenty seconds never felt longer. "In this format, you have an increase in your two energy systems: both the aerobic and the anaerobic," says Young. "Using your anaerobic system helps you build your endurance."

What It Feels Like The key to Tabata is pushing yourself to the wall of your peak performance at every interval—you know, that point at which you feel as if you're either going to pass out or pass up your lunch. Each exercise has modifications for optimal performance—during the push-up sequence, for example, I opted for run-of-the-mill push-ups, while the bodybuilder next to me clapped between them. You can easily adapt the workout by applying the 20-seconds-on, 10-seconds-rest model to your exercise of choice. But don't go overboard, Young warns. She recommends doing Tabata two to three times a week at most. "Your body will be too taxed to do it every day," she cautions.

PUMPING HIM UP *Gold medalist Michael Phelps used music as an essential part of his Olympic warm-ups.*

The Perfect Playlist

The tunes you put on your iPod will help you go faster and harder—even when you feel spent.

 BY ALEXANDRA SIFFERLIN

SOMETIMES YOU NEED AN EXtra push to hit the pavement or treadmill—or to make it through that last grueling mile of training—and the key may simply be loading the right songs on your iPod, according to Dr. Costas Karageorghis, co-author of *Inside Sport Psychology* and a leading expert on the psychophysical and ergogenic effects of music at Brunel University in London.

Music has motivational qualities that can make you work harder, even when your lungs are burning and your heart is beating at a high tempo. "Music has the propensity to elevate positive aspects of mood such as vigor and excitement, and reduces negative aspects such as tension and fatigue," says Karageorghis, who has created custom workout soundtracks for athletes who competed in the 2012 London Olympics. "It reduces perceived effort, and training to a musical beat can enhance endurance."

Whether you're a casual runner, a serious marathoner or an Olympian, the right playlist can optimize your performance. Here are Karageorghis's guidelines for putting together a runner's mix that will get you across the finish line.

Select Tracks With Energizing Beats Synchronizing your strides with an upbeat song can increase your effort during a workout. In a 2012 study, Karageorghis and his colleagues found that matching treadmill exercise with music significantly boosted exercise efficiency and endurance. The researchers studied 30 participants working out on a treadmill who listened to high-energy rock and pop tunes. Compared with those who worked out in silence, those who synchronized their gait with uptempo music improved their endurance by 15% over exercising without music.

Jamming to rhythmic songs also lowers your perceived exertion, making you think you're not working as hard as you really are. Upbeat music increases activity in a part of the brain called the ascending reticular activating system, which "psychs" you up.

The optimal tempo range for background music is 120 to 140 beats per minute, and the tempo can vary from there depending on the activity an individual wants music to sync to. "Our research shows this yields the best psychological outcomes," says Karageorghis.

By looking up the beats per minute (BPM) of your go-to songs, you can also find the tempo that matches the heart rate you want to achieve during your workout. For example, if you want your heart rate to get to 130 BPM, choose a song the tempo of which progressively increases to that beat, Karageorghis says.

Stick With What You Know A song's cultural impact is a key factor in what makes it motivational. "There's a strong relationship between exposure to a song and you liking it," says Karageorghis. We tend to favor songs the more often we hear them, so pick a song that's already in your music library.

Adding songs you associate with moments of perseverance, either from movies or your personal life, can also give you an extra edge. "[The] *Chariots of Fire* [theme song] was used extensively at the London Olympic Games," says Karageorghis. "We've made an association with this song and characters doing heroic feats. When you hear it, it conjures images and thoughts of overcoming adversity and striving toward a goal. So you're conditioned to feel stimulated, inspired and motivated."

One of TIME's staffers, photo editor Liz Ronk, says this strategy worked for her while training for a half-marathon: "Sometimes I hear songs that are played at races that I would normally never listen to, and I'll download them specifically for my runs just because the song will remind me of that energy."

Don't Forget to Hit Shuffle If you've had your playlist on repeat for the past two weeks, you may be desensitized to the songs' motivational qualities. "This is why radio stations promote a song by playing it repeatedly but then play it less and less, so listeners don't develop a negative response to it," says Karageorghis. "Change your playlist at least every couple of weeks so you don't listen to the same track over and over."

Try Some Digital Alteration To create playlists for professional athletes, including Britain's track-and-field captain, Dai Greene, Karageorghis films them working out at different intensities to identify tracks from their music libraries that fit their workouts. Then he tweaks the music to get them working even harder. "Often I digitally adjust tracks to give a little push of one or two beats per minute," says Karageorghis. "Differences in tempo of up to four beats per minute are indiscernible to nonmusicians. You can easily manipulate your favorite tracks slightly. It's a particularly good ploy if you want to give yourself a little jolt or get out of a training slump."

Be Choosy About Lyrics "Lyrics can be extremely important, particularly if they carry meaning for the athlete," says Karageorghis. "You will notice a lot of athletes, like Michael Phelps, use music as an integral part of their pre-event routine. He's famed for his rap-centric playlist. In Beijing [for the 2008 Olympics], he listened to the song 'I'm Me' by Lil Wayne, which has strong affirming lyrics as well as being acoustically stimulative." Find songs with inspiring lyrics that convey what you want to achieve, like "Pump It" by the Black Eyed Peas or "Lose Yourself" by Eminem.

Sample This...

If you're still unsure where to start, here are sample playlists from Karageorghis and from two recent half-marathoners-in-training:

DR. COSTAS KARAGEORGHIS

"Eye of the Tiger" (109 BPM), Survivor
"Don't Stop Me Now" (154 BPM), Queen
"Beat It" (139 BPM), Michael Jackson
"I Like to Move It" (123 BPM), Reel 2 Real
"Push It" (130 BPM), Salt-N-Pepa

HALF-MARATHON TRAINERS

"Available," The National
"Don't Save Us From the Flames," M83
"Ready to Start," Arcade Fire
"Dog Days Are Over," Florence + the Machine
"All of the Lights," Kanye West

"40 Day Dream," Edward Sharpe
   and the Magnetic Zeros
"Celebration Day," Led Zeppelin
"Paper Planes," M.I.A.
"No Regrets," Aesop Rock
"I Can't Turn You Loose," Otis Redding

Goal Power

Deciding to be healthy isn't enough. Fortunately, we're learning the secrets of turning resolution into action. My guide to getting unstuck.

✳ BY DR. MEHMET OZ

AM TERRIFIED OF HEIGHTS. CLIMBING TO THE TOP of a telephone pole three stories tall in the Arizona desert is not my idea of a good time. It should have helped that I'd be wearing a helmet and a safety cable, but that made no difference. If you know anything about how the brain works, you know that what you understand in your reasoning lobes and what you feel in your emotional lobes are two very different things. And when it came to the idea of standing atop a swaying, creaking pole with the desert floor swimming below, my emotional lobes won in a landslide.

I had gone to Arizona with a crew from *The Dr. Oz Show* and a group of 50 women who had been offered a surprise trip because they desperately wanted to make changes in their lives. Maybe one wanted to quit smoking; maybe another wanted to lose weight or to exercise more or to get out of a lousy job. The point is that they were stuck, rooted between the world of "I know I should do this" and the world of "I'm actually doing it," a place where many of us can spend a lot of fruitless and frustrating time.

The three-story pole was a three-story metaphor. Take a step forward—or actually, a step straight up—in Arizona and maybe we could all take other, even scarier steps elsewhere in our lives. I was paired with Pam, a 33-year-old IT worker who had been obese as a child and hasn't wanted to be singled out since, for fear of embarrassment. So in the spirit of taking the counterintuitive step, she agreed to go first and be singled out for a very positive reason. I

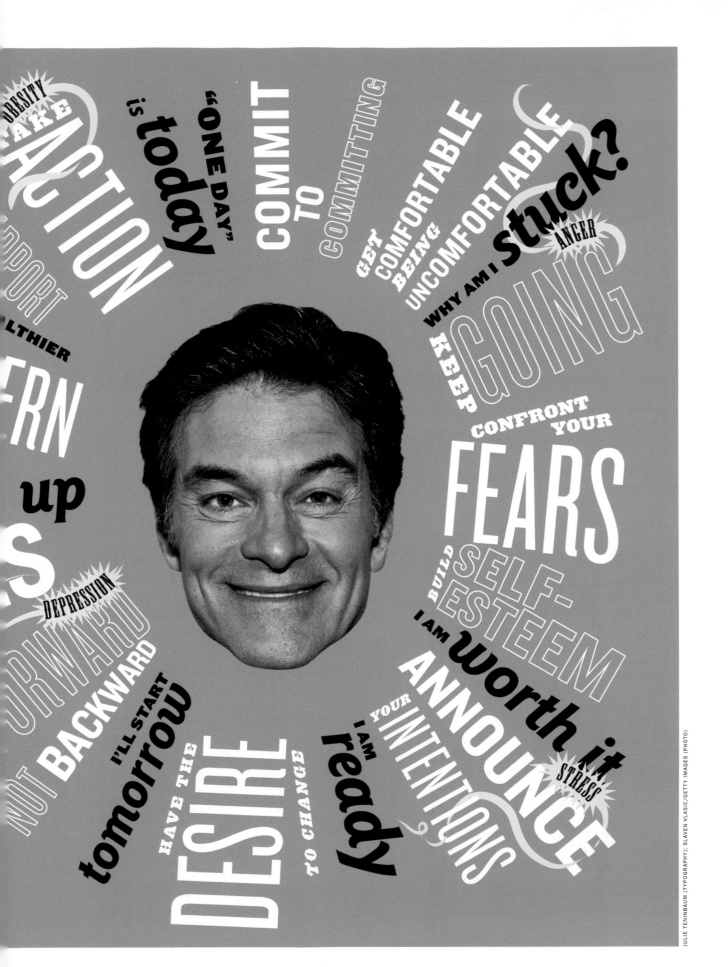

stood at the bottom of the pole, watching her go up and dreading the moment when she would reach the tiny platform at the top, which was big enough for only two, because that would mean I had to go. When I began the climb, it was just what I expected, by which I mean the whipping wind and the receding ground and my sweaty palms on the rungs were just awful. But there was a power in enduring that awfulness.

Pam shouted down encouragement, and as I neared the little platform, she reached out, took my hand, and helped me take the final step. My real fear, it seemed, had been not of heights but of lack of control. Trusting Pam was the antidote. I flung an arm around her in a celebratory hug. The climb may have been terrible, but completing it was sublime.

You Know You Should

There's a strange, discordant pathology to the state of being stuck. Nobody who starts smoking plans to be hooked for life. Nobody who's obese secretly wants to stay that way. When a doctor or family member pleads with us to make lifesaving changes, we mean it when we say, "I know, I know, I should. I will." What's left unsaid is the killer caveat: "Just not today."

Over the years, I have had more conversations like this with patients than I can count, so many that the phrase "I know I should" has become a bright red flag—a sad predictor that I will probably one day crack open those patients' sternums in the operating room. Why these people spend their lives moving stubbornly toward major, potentially life-threatening surgery when they could avoid it with just a few wise moves is a riddle that has always dogged me.

But perhaps it shouldn't. Getting people to make meaningful changes in their lives is much more complicated than explaining to them what to eat for dinner, how often to exercise, and which kinds of tests they should get from their doctors. The psychology of health is every bit as complex as the biology.

A colleague and frequent collaborator of mine, psychologist John Norcross, is one of the authors of a groundbreaking book, *Changing for Good*, that focuses on the transtheoretical model, which has been around since the late 1970s but is no less powerful now. This model divides the process of making changes into five key stages: precontemplation, contemplation, preparation, action and maintenance. Precontemplation means you have no plans to make any changes and may not even be aware that something is wrong. This is often the state of the early-stage alcoholic who clearly has a problem but has not yet lost a job or cracked up a car and can thus pretend everything's fine. It's less common in cases of obesity or smoking. There are too many reminders, from the warnings on cigarette packs to the pains in your chest to the numbers on the scale. Much trickier are the contemplation stage, in which you begin to recognize the problem and debate taking action, and the preparation stage, when you intend to act imminently. People who get stuck at "I know I should" are generally marooned on the shores of one of these states of mind.

So how to get unstuck? Here's where psychology and a little spirituality can help. Throughout time, religion has been about not just worship but also life lessons, self-improvement and redemption, with earthly accountability to the community and congregation to help keep us in line. It's hard to cheat a neighbor who sits next to you in church. It's hard to skip an exercise class organized by your congregation. Alcoholics Anonymous was launched in the 1930s with a 12-step model based on the same idea. Here, too, getting the support—and if necessary, the scolding—of a group leads to better results than relying on sheer teeth-gritting willpower.

A 2012 study in the journal *Obesity* explored the power of the collective, looking at the weight-loss rate among people in groups who succeeded at a 12-week diet. The 12,000 participants divided themselves into 987 five- to 11-person teams, stayed in regular communication over the Internet, and competed with other teams to see who could lose the most at the end of 12 weeks. In general, the researchers found there was a kind of virtuous loop among the most successful groups: the more pounds any individual member of a high-weight-loss group dropped, the likelier it was that other members would shed more pounds too. At the end of the study, members of the most motivated and successful groups lost 5% of their body weight—a healthy and maintainable target that members of many of the other groups missed. Crunching the data, the researchers concluded that any person from a 5% group would have lost only 3.8% of body weight if moved to a less successful group. In other words, it wasn't just the quality of the individual's resolve; the quality of the group's resolve mattered too.

A 2008 study published in the *New England Journal of Medicine* took a similar approach, looking at the dynamics of smoking in a large social network. Harvard's Dr. Nicholas Christakis, the lead author of the paper, found that a whole range of health issues, from depression to smoking to obesity, can be powerfully influenced by the other people in your social web—including some you've never met. There's no telepathy involved in that. Perhaps you decided to quit smoking because you saw a friend succeed at it. But she quit because her husband, whom you know slightly, had done so. And he quit because he was following the example of an office friend you don't know at all. Twang one strand in the web and it can cause vibrations everywhere. The 2008 study explored more direct, first-person contact, but the results were still striking: people whose spouse quit smoking were 67% less likely to start or continue smoking; they were 36% less likely if a friend quit, 34% if a work colleague quit, and 25% if a sibling did.

Achieving a goal doesn't always have to be a collec-

1. PRECONTEMPLATION   2. CONTEMPLATION   3. PREPARATION   4. ACTION   5. MAINTENANCE

Change Starts Here

The transtheoretical model's five steps to getting unstuck:

1. Precontemplation
You're not even thinking about change. In some cases, you don't know a problem exists.

2. Contemplation
This is when you've begun considering the pros and cons of changing but you've made no decision.

3. Preparation
Change could be imminent. You begin taking proactive steps like telling family and friends.

4. Action
A behavior has been changed—which can be thrilling at first but requires commitment to continue.

5. Maintenance
The new behavior has been practiced for six months. Slips are still a risk.

tive effort. Sometimes it can be simply a matter of breaking the job up into a lot of minigoals. In a 2007 randomized clinical trial for postmenopausal women with Type 2 diabetes, investigators tested the effectiveness of the Mediterranean lifestyle, a regimen of regular exercise and a diet based on fruits, vegetables, olive oil, nuts, beans and lean meats, particularly seafood. Subjects who set short-term goals—increasing exercise sessions in five-minute increments, for instance—were doing better at the end of two years than a control group of people who simply adopted the program and, all too often, abandoned it later.

The Accountability Factor
Medication compliance is another critical area that can be improved by intervention and goal setting. From 50% to 75% of all hypertension patients currently prescribed drugs don't have their blood pressure under control because they fail to take the meds dependably. A 2006 study conducted by the National Heart, Lung and Blood Institute and published in the journal *Disease Management* used the transtheoretical model to try to improve those numbers in a group of 1,227 patients. All the subjects were in the pre-action phase: not complying but planning to. And all were given the assistance of both an on-screen and a paper workbook that used questionnaires and other tools to explore why they were stuck and what it would take to get them to move. Significantly, participants were surveyed three times during the study, meaning there was follow-up and accountability in the mix. A year after the study was done, 73% of those who had received counseling were compliant with their drug regimens, compared with just under 58% for a control group. At 18 months it was 69% to 59%. That's not 100%, but imagine slowing the No. 1 killer in the U.S. by simply getting resistant patients to take their statins and beta blockers.

I'm not pretending any of this is easy. Inaction is a powerful coping mechanism. Justification and rationalization are ways to handle feeling desperate and out of options. The National Institute of Mental Health published a revealing article in 2010 on the phenomenon known as emotional inertia—a sort of fixed state of depression, low self-esteem, anxiety or other condition that rarely seems to change, even in the face of circumstances that warrant change. I see this in some of my patients, who are so demoralized by their failure to lose weight or quit smoking that as soon as they leave my office, they seek comfort in a sundae or a cigarette.

In fairness, that is the nature of all of us. Human beings can be remarkably static creatures; it's practically woven into our DNA. Why move from the familiarity of the campfire circle and step into the scary wilderness, even if wonderful, life-giving things might be there? Taking action, of course, is the very hardest of the transtheoretical steps—the moment you stand up from the campfire, dust off your pants and begin walking toward the woods. Difficult as it is, though, it can be the start of an inexorable forward momentum. As Newton taught, a body in motion stays in motion. I often remind my patients of a favorite and familiar riddle: If you see 10 birds on a wire and five decide to fly, how many are left? The answer is 10. Deciding and doing are not the same.

I learned that fact anew at the base of that pole in Arizona. I traveled west that day having already decided to climb the blasted thing. But I hadn't done anything at all until I actually put my foot on the lowest rung.

About the Authors

DAVID BJERKLIE is a science writer and the author of children's books on butterflies, agriculture and environmental justice. Formerly he was the senior science reporter at TIME, a senior editor at TIME for Kids and a science writer/editor at TheVisualMD.com.

JOHN CLOUD is a former staff writer for TIME who wrote dozens of in-depth features, including cover stories on organic food, gay teenagers, and such diverse figures as Ann Coulter and Howard Dean.

STACEY COLINO writes about health, wellness and beauty for *Real Simple, More, Parents* and other national magazines. She is the co-author, with Dr. David Katz, of *Disease-Proof: The Remarkable Truth About What Makes Us Well.*

KRISTINA DELL is a freelance writer in New York who writes about business, education and law.

CHRISTINE GORMAN is the health/medicine/biology features editor at *Scientific American.*

JEFFREY KLUGER is a senior editor at TIME, overseeing its science and technology reporting. He has written or co-written more than 40 cover stories for the magazine and regularly contributes articles and commentary on science, behavior and health.

KRISTIN KOCH is a writer, editor and lifestyle expert based in New York.

MICHAEL D. LEMONICK is a senior science writer at the research organization Climate Central. He covered science for TIME magazine for more than 20 years and is the author of several books on astrophysics, including *Echo of the Big Bang; Other Worlds: The Search for Life in the Universe;* and *Mirror Earth: The Search for Our Planet's Twin.*

CATHERINE MAYER is TIME's Europe editor. She is the author of *Amortality: The Pleasures and Perils of Living Agelessly.*

HARRY MCCRACKEN is an editor at large at TIME, where he writes about personal technology.

COLLEEN MORIARTY is a Connecticut-based freelance magazine writer and author of *Shortcuts to Sexy Abs: 337 Ways to Trim, Tone, Camouflage and Beautify.*

MYATT MURPHY has written about exercise, nutrition and health for over 20 years and is the author of eight books, including *The Body You Want in the Time You Have, Men's Health Ultimate Dumbbell Guide* and *Testosterone Transformation.*

REGINA NUZZO has a doctorate in biostatistics and writes about science and health for *Reader's Digest, Nature News* and other publications.

MEHMET OZ, M.D., is a vice chairman and professor of surgery at Columbia University, a bestselling author and the Emmy-winning host of the *The Dr. Oz Show.*

ALICE PARK is a staff writer at TIME. She has reported on the frontiers of health and medicine in articles about such issues as cancer and genetics.

JULIA SAVACOOL is the author of *The World Has Curves.* She writes for numerous publications, including *Health, Self, Glamour, Details, USA Today* and ESPN.com. She lives in New York City.

ALEXANDRA SIFFERLIN writes about health for Time.com. She is a graduate of Northwestern University's Medill School of Journalism.

JOEL STEIN is a columnist for TIME. His first book, *Man Made: A Stupid Quest for Masculinity,* is in bookstores now.

MAIA SZALAVITZ is a neuroscience journalist for Time.com and co-author of *Born for Love: Why Empathy Is Essential—and Endangered.*

BRYAN WALSH is a senior editor for TIME International. He also writes the "Going Green" column for TIME and Time.com and contributes to that website's environmental-issues blog, "Ecocentric."

MARTHA C. WHITE writes about consumer credit, debt and retail banking for Time.com and previously contributed to AOL's WalletPop.com. She has written about business, finance and the economy for outlets including Slate, the *New York Times,* MSNBC.com and *Fast Company.*